Immobilized Cells and Organelles

Volume II

Editor
Bo Mattiasson, Ph.D.
Associate Professor
Department of Pure and Applied Biochemistry
Chemical Center
University of Lund
Lund, Sweden

CRC Press
Taylor & Francis Group
Boca Raton London New York

CRC Press is an imprint of the
Taylor & Francis Group, an **informa** business

First published 1983 by CRC Press
Taylor & Francis Group
6000 Broken Sound Parkway NW, Suite 300
Boca Raton, FL 33487-2742

Reissued 2018 by CRC Press

Library of Congress Cataloging in Publication Data
Main entry under title:

Immobilized cells and organelles.

Bibliography: p.
Includes index.
Immobilized cells. 2. Immobilized cell organelles.
I. Mattiasson, Bo, 1945-
QH585.5.I45I47 1982 547.87 83-17870
ISBN 0-8493-6440-X (v. 1)
ISBN 0-8493-6441-8 (v. 2)

A Library of Congress record exists under LC control number: 82017870

Publisher's Note
The publisher has gone to great lengths to ensure the quality of this reprint but points out that some imperfections in the original copies may be apparent.

Disclaimer
The publisher has made every effort to trace copyright holders and welcomes correspondence from those they have been unable to contact.

ISBN 13: 978-1-315-89428-7 (hbk)
ISBN 13: 978-1-351-07338-7 (ebk)

Visit the Taylor & Francis Web site at http://www.taylorandfrancis.com and the
CRC Press Web site at http://www.crcpress.com

PREFACE

Besides giving some insight into basic technology (immobilization procedures, etc.) these volumes also sum up the current know-how in the area and try to predict some future trends.

Cells and organelles are small units for biochemical synthetic purposes, often the smallest practically feasible unit since they contain coenzyme regenerating system, ordered enzyme sequences, etc.

The term "immobilized cells" covers everything from dead cells with a single active enzyme species to cells proliferating on or within a three dimensional polymer matrix. The practical handling of these structures make them useful in various applications, e.g., large-scale production of biomolecules, biodegration, analysis, etc.

In recent years immobilization techniques have become very mild so that, besides microorganisms, plant and animal cells can also be immobilized. When immobilized enzymes first appeared a bright future for the technology was predicted. During the last years some severe limitations have appeared, e.g., the problems to develop a practically feasible technology for coenzyme regeneration, the lack of methods to arrange enzyme molecules in ordered clusters to perform multistep enzyme catalyzed reactions. To all these problems immobilized cells and organelles offer exciting and promising alternatives.

Is it possible to manipulate the total metabolism by immobilization of the cell? This question is not fully answered yet, but strong evidence shows that changes occur.

Immobilized animal cells grown on microparticles of weak ion-exchangers have changed the field of cell-culture dramatically. This technique now makes it possible to exploit mammalian cell lines in biotechnology.

Bo Mattiasson

THE EDITOR

Bo Mattiasson, Ph.D., is an Associate Professor in the Department of Pure and Applied Biochemistry, Chemical Center University of Lund, Lund, Sweden. Dr. Mattiasson received his Ph.D. degree in biochemistry from the University of Lund in 1974. He has been teaching basic and applied biochemistry since 1971, and since 1976 has been an Associate Professor in biochemistry.

CONTRIBUTORS

Peter Brodelius, Ph.D.
Professor
Department of Pure and Applied Biochemistry
University of Lund
Chemical Center
Lund, Sweden

Julian Clark, Ph.D.
Technical Development Manager
Cell Biology Group
Pharmacia Fine Chemicals AB
Uppsala, Sweden

S. O. Enfors, Ph.D.
Department of Biochemistry and Biotechnology
Royal Institute of Technology
Stockholm, Sweden

Saburo Fukui, Ph.D.
Professor
Department of Industrial Chemistry
Faculty of Engineering
Kyoto University
Kyoto, Japan

Stina Gestrelius, Ph.D.
Section Head
Immobilized and Derivatized Enzymes R & D
Novo Industria A/S
Bagsvaerd, Denmark

Lena Häggström, Ph.D.
Department of Biochemistry and Biotechnology
Royal Institute of Technology
Stockholm, Sweden

Bärbel Hahn-Hägerdal, Ph.D.
Technical Microbiology
Chemical Center
University of Lund
Lund, Sweden

Michael D. Hirtenstein, Ph.D.
Product Group Manager
Cell Biology Group
Pharmacia Fine Chemicals AB
Uppsala, Sweden

John Hopkinson, Ph.D.
Director
New Business Development
Amicon Corporation
Danvers, Massachusetts

S. B. Karkare
Department of Chemical and Biochemical Engineering
Rutgers University
New Brunswick, New Jersey

Bo Mattiasson, Ph.D.
Associate Professor
Department of Pure and Applied Biochemistry
Chemical Center
University of Lund
Lund, Sweden

Jose-Luis Ochoa, Ph.D.
Biotechnology Program
Centro de Investigaciones Biologicas
La Paz, Mexico

Atsuo Tanaka, Ph.D.
Senior Lecturer
Department of Industrial Chemistry
Faculty of Engineering
Kyoto University
Kyoto, Japan

K. Venkatasubramanian, Ph.D.
Director, Corporate Research
Technical Development
H. J. Heinz Company
Pittsburgh, Pennsylvania
Visiting Associate Professor
Rutgers University
New Brunswick, New Jersey

TABLE OF CONTENTS

Volume I

Volume II

Chapter 1

IMMOBILIZED NONVIABLE CELLS FOR USE OF A SINGLE OR A FEW ENZYME STEPS

Stina Gestrelius

TABLE OF CONTENTS

I. INTRODUCTION

For a process with only one or a few reaction steps the biocatalyst of choice may be a single enzyme, a couple of enzymes, or a cell containing the relevant enzyme(s). To date most industrial biocatalyzed processes are carried out either with soluble, cheap preparations of extracellular enzymes (as e.g., proteolytic, amylolytic or lipolytic enzymes) or by conventional fermentations (steroid conversions, production of gluconic acid, vinegar, etc.).

Very few intracellular enzymes are produced and used industrially, mainly due to severe problems with large-scale isolation procedures.[1] Especially membrane-bound enzymes are often solubilized and recovered in too low yields. For utilization of intracellular enzymes from microbial cells it has therefore been a natural approach to use the whole cell material, and a similar development can be expected for plant enzymes (see Chapter 3, Volume I).

In the early industrial applications the cells were used once only. However, by simple flocculation[2] or by choosing inherently stable spores[3] or fungal pellets[4] the cells could be used repeatedly.

By further physical and chemical treatment leading to a preparation of immobilized nonviable cells, the operational stability has often been improved, thus permitting re-use during increased periods of time (months) while the improved physical stability has allowed a wider choice of reactors and continuous processing. Yet the number of immobilized cell systems, viable or nonviable, that are presently used industrially is very modest. Table 1 lists such commercially operating immobilized nonviable cell catalysts, and it can be noted that the list would not be much longer even if immobilized cell-free enzymes or immobilized viable cells were included. However, except for processes like glucose isomerization where immobilized preparations of enzymes or cells are without competition, the introduction of particulate biocatalysts must be expected to occur only gradually when old plants are replaced, to fulfill new requirements of low energy consumption and increased in-house and environmental safety.

This paper will primarily deal with the interdisciplinary efforts required in developing new biocatalysts of the type immobilized nonviable cells for medium-size or large-scale applications where the keywords are high volumetric activity and good operational stability during continuous processing.

II. DEFINITION

An immobilized nonviable cell preparation contains, by the present definition, all or almost all of the original cell components, but may have lost more or less of the original cell organization. The cell structure may have become impaired before, during, or after immobilization, either by physical treatment such as homogenization, lysis, heat-treatment or freeze-thawing, or by treatment with solvents, detergents, or other chemicals.

III. RATIONALE FOR IMMOBILIZING NONVIABLE CELLS

In addition to the producer's advantage of immobilizing cells when the desirable enzyme(s) cannot be economically isolated or stabilized in a cell-free form, there must also be one or several technical advantages if the application of a nonviable cell biocatalyst is to become successful. Examples of such advantages include simplified catalyst re-use, smaller reactors, continuous processing, improved catalyst stability, withdrawal of catalyst from product stream, less unit operations in the product upgrading, and cheaper waste treatment.

Table 1
COMMERCIAL IMMOBILIZED NONVIABLE CELL PREPARATIONS

Utilized enzyme	Application	Microorganism	Immobilization method	Producer (trade name)	Ref.
Aspartase (E.C. 4.3.1.1)	Production of L-Asp. from ammonium fumarate	*E. coli*	Entrapment in K-carrageenan, treatment with hexamethylenediamine + glutaraldehyde	Tanabe Seiyaku Co. Ltd., Japan	5
Fumarase (E.C. 4.2.1.2)	Production of L-malic acid from fumaric acid	*Brevibacterium flavum*	Entrapment in K-carrageenan	Tanabe Seiyaku Co. Ltd., Japan	6
Alpha-Galactosidase (E.C. 3.2.1.22)	Hydrolysis of raffinose (in molasses) to galactose and sucrose	*Absidia regneiri*	Cross-linking after freeze-thawing	Nippon Beet Sugar Mfg. Co. Ltd. Japan	7
		Mortierella vinacea	Cross-linking with glutaraldehyde	The Great Western Sugar Co., U.S.A.,	8
Beta-Galactosidase (E.C. 3.2.1.23)	Hydrolysis of lactose (in whey) to glucose + galactose	*Bacillus sp.*	Cross-linking with glutaraldehyde	NOVO Industri, Denmark (Novozym 231)	9
Glucose isomerase (E.C. 5.3.1.5)	Production of high fructose syrup from glucose	*Actinoplanes missouriensis*	Occlusion in gelatine, cross-linking with glutaraldehyde	Gist-Brocades, Netherlands (Maxazym)	10
		Bacillus coagulans	Cross-linking with glutaraldehyde	NOVO Industri, Denmark (Sweetzyme)	11
		Streptomyces olivacens	Cross-linking with glutaraldehyde	Miles Laboratories Inc., U.S.A. (Taka Sweet)	12
		Streptomyces albus	Heat treatment Filter reactor	Clinton Corn Proc. Co., U.S.A.	13
Invertase (E.C. 3.2.1.26)	Hydrolysis of sucrose to fructose and glucose	Selected yeast strain	Occlusion in gelatine, cross-linking with glutaraldehyde	Gist-Brocades, the Netherlands (Maxinvert Immob.)	14
Penicillin acylase (E.C. 3.5.1.11)	Production of 6-APA from penicillin G or V	*E. coli*	Entrapment in polyacrylamide	Tanabe Seiyaku Co. Ltd., Japan	15
		Proteus rettgeri	Binding with glutaraldehyde to glycidylmethacrylate polymer	Pfizer Inc., U.S.A.	16
		Selected microorganism	Cross-linking with glutaraldehyde	NOVO Industri, Denmark (Novozym 217)	17
		E. coli	Entrapment in gelatine, cross-linking with glutaraldehyde	Shanghai, China	18

A number of limitations have also been pointed out, such as difficulties with high molecular weight reactants where soluble enzymes are generally preferred, and difficulties with cofactor retention and regeneration where viable cells may be preferred. Finally such applications, e.g., in medical treatment, where a well-defined catalyst composition is required and a pure (immobilized) enzyme is regarded safer.

For single enzymes without need for cofactors, the immobilization of *viable* cells is chosen mainly to secure a long lifetime of the catalyst through intermittent reactivation in suitable medium (see Chapter 2, Volume II). However, the application of living cells necessitates the use of very mild immobilization procedures and virtually aseptic application conditions to preclude microbial contamination during reactivation, making this type of catalyst attractive for only few industrial branches. *Nonviable* cells, on the other hand, can be treated with heat and chemicals to the extent that the enzyme of interest can withstand, permitting the use of a larger number of immobilization methods and sterilization treatments. Other advantages include facilitated mass transport to and from the enzymes after rupturing or permeabilizing the cells, and the potential of inactivating disturbing enzymes to obtain cleaner reactions and improved operational stability (the latter after removing proteolytic enzymes).

Thus, for many single enzyme applications there is a potential of finding a combination of microorganism strain and immobilization procedure that will permit production of a cheap immobilized nonviable cell catalyst with as high volumetric activity and clean reaction pattern as most immobilized enzyme preparations. Also for some multienzyme applications, e.g., of consecutively working enzymes without cofactor requirements, the concept of immobilizing one type of nonviable cells containing all the desired enzymes may prove very favorable.

IV. PRODUCTION OF IMMOBILIZED NONVIABLE CELLS

A. Selection of Cells

Like in all enzyme production a screening program is normally undertaken in order to find the best enzyme source for the application of interest.[19]

Special emphasis must, however, be put on selecting a cell that is suitable not only for producing the best enzyme, but also for becoming an integrated part of the immobilized preparation. A necessary condition is, of course, that the enzyme of interest is intracellular. While a high enzyme activity per cell is a generally desirable criterium that can often be met by mutational improvement of a strain, the stability during immobilization and the subsequent performance under application conditions are properties that are difficult both to predict and correct.

Takata et al.[20] have described a screening program for K-carrageenan immobilized fumarase where 4 strains out of 241 were chosen for thorough investigation of immobilization yields and operational stabilities of the immobilized cells. Each of the four strains *(Brevibacterium ammoniagenes, Brevibacterium flavum, Proteus vulgaris,* and *Pseudomonas fluorescens)* was fermented on a separately optimized medium, and the strain that turned out the best choice for catalyst activity and stability was not the one with the highest activity on screening medium. Thus, even with a prechosen immobilization method the selection of the most favorable strain is very time-consuming. When both the cell strain and the immobilization method are to be selected for developing a new nonviable cell catalyst, an iterative approach may be useful[21] since test runs with a suboptimal catalyst may help outlining the evaluation criteria. Safety regulations for production and application of microbial cell material (especially food industry regulations) require that microbial strains and fermentation media are selected with great care.[22] Pathogenicity and toxicity testing is thus becoming a routine procedure in the industrial strain selection programs. The choice of asporogenic organisms or mutants may reduce viable counts of enzyme/nonviable cell preparations and increase safety in production.[19]

The absence of interfering enzyme activities is a very important property of cells to be immobilized. Yamamoto et al.[23] thus selected a *Pseudomonas putida* (L-arginine deiminase) for citrulline production since this strain lacked the undesirable ornithine transcarbamylase, and Jack and Zajic[24] chose a *Micrococcus luteus* strain that was free of urocanase for urocanic acid production with His-ammonia lyase.

Permeabilizing treatment causing loss of cofactors has proved effective in removing side activities by cofactor-dependent enzymes. A classical example is bile extract treatment of *Brevibacterium ammoniagenes* (fumarase catalyst) as described by Yamamoto et al.,[25] leading to suppression of unwanted succinic acid formation. Cell rupture treatments in connection with immobilization will be discussed in detail in Section IV.C. below (for references see Table 4).

Heat and pH shocking can sometimes be a selective method for removing hydrolytic enzymes,[26-28] but a selected strain or mutant lacking the interfering enzyme(s) offers the most convenient and permanent solution to the side activity problem.

B. Production and Recovery of Cells

After the microorganism and strain have been selected, and maybe improved, much attention must be paid to the properties of the fermentation broth since it will strongly influence the quality of the final immobilized cell preparation. For example the choice of raw materials, sterilization method, and foam control method may affect not only the yield and activity of the catalyst, but also the operational and physical stability.

Cheap media often have high contents of solids such as ground whole grains or flakes which may be difficult to separate from the cell sludge prior to immobilization. It should be remembered that the dry cell yield in a typical large-scale enzyme fermentation is often 0.5 to 2% while residual nutrients and metabolites constitute 5 to 10% of the broth at the end of a fermentation.[19] The cells are generally recovered from the broth by centrifugation or filtration prior to immobilization, but some immobilization methods comprising flocculation,[29] chelation,[30] or cross-linking[31] as initial steps can also be performed without previous separation. Such procedures can be favorable if the cells are difficult to recover.

For the manufacture of carrageenan-immobilized *Brevibacterium flavum*/fumarase (catalyst for malate production), Takata et al.[20] have established the optimum culture conditions including carbon source, nitrogen source, vitamins, effect of corn steep liquor, aeration rate, temperature, and time. Maximum operational stability of the immobilized cells was found for the same culture condition as maximum activity of free (and immobilized) cells with a few exceptions. A change in the concentration of corn steep liquor had only effect on activity while changes in the aeration rate had the most pronounced effect on the operational stability.

Fermentation time can sometimes be a very important parameter for activity yield and stability of immobilized cell preparations, e.g., for cells that easily lyse and leak intracellular enzymes if they are not harvested at the proper stage. The choice of certain cells and fermentation techniques may simplify immobilization. It has, for example, been proposed to use autoflocculating cells or marine bacteria that have an inherent tendency to attach to surfaces.[32] Kennedy et al.[33] found it easy to chelate a cellulosic slime-producing *Acetobacter* strain with hydrous titanium oxide while nonflocculating strains could not be chelated by the same method. Also the induction of filamentous growth of bacteria has been used to facilitate cell immobilization.[34]

The good inherent stabilities of spores and conidia have been utilized for preparing immobilized cell catalysts by adsorption,[35] entrapment, or mixing with filter aid.[36] (Entrapment of spores followed by germinating treatment has recently become a valuable method for aseptical production of viable cell catalysts[37,38]). Finally, mycelium pellets have been transformed into stable biocatalyst particles by cross-linking with glutaraldehyde.[39]

C. Selection of Immobilization Method

While hundreds of cell-immobilization methods have been described, including en-

Table 2
DESIGN CRITERIA FOR INDUSTRIAL IMMOBILIZED NONVIABLE CELL PREPARATIONS FOR USE OF A SINGLE OR A FEW ENZYMES

Item	Requirement of biocatalyst or immobilization method
Application (e.g., food; pharmaceuticals; organic synthesis; waste treatment)	Toxicity testing. No leakage. Choice of harmless reagents. Sufficient productivity.
Reactor (e.g., type: fixed-bed, fluid-bed, stirred tank; size)	Physical stability. Suitable density, size, shape. Sufficient volumetric activity.
Reaction conditions (e.g., pH; temperature; solvents)	Chemical resistance, thermostability. Hydrophilicity/hydrophobicity.
Reactant properties (e.g., molecular weight, purity, solubility or volatility of substrates and products)	Suitable pore size and particle size. Elasticity.
Cell type (e.g., unicellar organism; cell size; robustness)	Suitable porosity of carrier. Rupturing treatment of cells.
Enzyme properties (e.g., location in cell; sensitivity to oxygen or chemicals; cofactor requirement)	Rupturing treatment or mild immobilization. Utilization of several enzymes (for cofactor regeneration)
Catalyst manufacture (e.g., scale; continuous/discontinuous, safety requirements, acceptable cost)	Simple upscaling. Cheap and nontoxic reagents. High yield. Good storage stability. Suitable form for transportation.

trapment, aggregation, adsorption, cross-linking, and chemical coupling to carriers (see Chapter 2, Volume I), the guidelines for selecting a suitable method are seldomly discussed. Table 2 lists a number of criteria that are important to consider in designing an industrial biocatalyst of immobilized nonviable cell type.

1. Catalyst Application

Although it may be tempting to judge a preparation from its activity and immobilization yield, it is generally safer to start from the other end and make sure that the method and preparation will be compatible with the potential application.

Even if immobilized biocatalysts will qualify as processing aids and not as additives in food or drugs, new product approval laws demanding extended toxicity testings are presently being passed in many countries.[40] Support materials, such as edible proteins and polysaccharides, and chemical reagents that are already being used in food processing (e.g., as cleaning agents) may then be more easily accepted than synthetic polymers and new chemicals.

In this context any leakage from a preparation is essential to control. Any leakage of cells will, of course, lead to decreased catalyst activity, but loss of cell or carrier material can also be serious in polluting reaction mixtures or cause problems in downstream processing. Entrapment and covalent coupling techniques are therefore often preferred to physical adsorption or aggregation.

For a certain industrial application the immobilized cell catalyst will be judged from its productivity in a particular reactor. The productivity concept denotes the accumulated amount of product formed per unit catalyst (kg or liter) during the operational lifetime. This lifetime is, of course, a compromise between production rate and catalyst cost, but two half-lives, i.e., until the catalyst has lost 75% of its original activity, is often the minimum time. For most industrial applications using fixed-bed reactors the useful lifetime should be in the order of months. This can, however, seldomly be attained unless microbial contamination can be avoided (contaminations will plug porous catalysts, destroy flow patterns, change pH, or decompose reactants or the catalyst itself), and it is therefore very important that the catalyst can be used under extreme conditions or can tolerate the addition of preservatives or intermittent washing procedures.

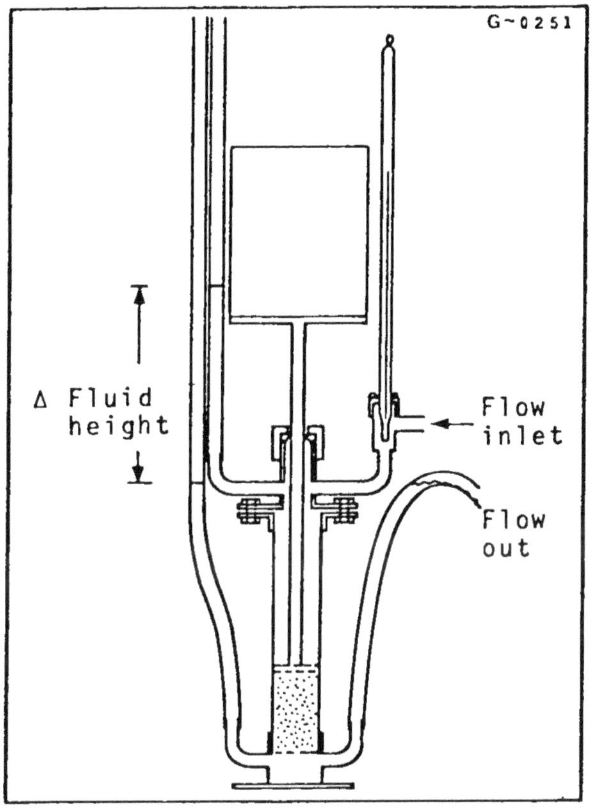

FIGURE 1. Pressure test apparatus. (From Norsker, O., Gibson, K., and Zitton, L., *Starch/Starke,* 31, 13, 1979. With permission.)

2. Reactor Type

Even if the reactor in principle can be chosen to suit a certain type of catalyst, as e.g., a spiral-wound biocatalytic reactor module for membranes[41] or a radial reactor for fibers,[42] it seems to be more common that the reactor design depends on industrial practice and investment costs. Packed beds are used in most large-scale applications due to a high volumetric efficiency and ease of operation. Where pH or temperature control are critical, an arrangement of rapid recycling[43] or a series of smaller packed beds with intermittent adjustment[17] can approach the stirred tank performance without the abrasion problem often seen in stirred tank systems. Batch-run or continuous stirred tanks are, however, also used, especially for penicillin acylase applications.[44]

Dependent on bed height as well as feed stream velocity and viscosity, the application of a biocatalyst in a fixed-bed reactor will require a certain physical stability. While glucose isomerase preparations are loaded into 5 to 7 m beds, a full-size application of a steroid-converting biocatalyst may require less than 0.5 m beds. Figure 1 shows a laboratory scale pressure test apparatus that has been developed by Norsker et al.[45] to simulate the behavior of particles in a large fixed-bed reactor. The apparatus measures the permeability of the true substrate through a sample of catalyst as a function of time at different compression forces, and it is used by NOVO Industri, Denmark, both for testing experimental products in the development stage and for routine quality control of commercial preparations. Similar set-ups have been described by Cheetham[46] and Klein and Kluge.[47] Compression tests on polymer blocks or single particles[48] give rapid estimations of pressure stabilities, but cannot be regarded as useful methods for testing the behavior of particles in a reactor.[49]

Stirred tank reactors and fluidized beds can be expected to subject an immobilized cell preparation to even harsher treatment than fixed-bed reactors. Klein and Eng[50] have described a method for measuring the abrasion of biocatalyst particles in a well-stirred vessel by following particle scattered Tyndall light with a nephelometer.

The many different test methods and reaction conditions applied makes it very difficult to use published data to compare biocatalyst particles. However, there seems to be a common finding for gel-entrapped preparations that both compression and abrasion resistance of particles decreases with increasing cell loading. High cell loading is not necessarily an advantage for fixed-bed reactors where the increased flow rate (to obtain unchanged substrate conversion) will require *further* improvement of the pressure stability of the particles. On the other hand, a maximum loading is desirable for stirred tanks since it leads to an increased volumetric activity and lowered volume fraction of the catalyst ($V_{particle}/V_{total}$), resulting in less contact between particles and stirrer and consequently less abrasion.[50] A similar difference between requirements for fixed bed and stirred tank reactors is also found for particle size since large particles are generally more pressure stable, but more easily abraised than smaller particles.

Several recent papers and patents have dealt with the improvement of particle strength. Thus, addition of small pre-formed particles[51] or polyamine polymers[52] during cross-linking of cells, hardening treatment of polysaccharide gel preparations with polyamines[53-54] or polyamine plus glutaraldehyde,[55] and drying of polysaccharide gels[56,57] are examples on procedures that have resulted in better physical stability of the preparations.

3. Enzyme Reaction Conditions and Reactant Properties

For a number of applications the processing parameters such as substrate concentration and purity, pH, temperature, presence and concentration of buffers or solvents, etc., are completely or partially fixed regardless of the properties of the biocatalyst.

Many examples are found in food processing (e.g., milk, beer, juice, wine), but also pharmaceutical production may be restricted by a lack of stability or solubility of the reactants. Thus, the application of penicillin acylase for 6-APA production is limited to temperatures below 40°C, a narrow pH range, and a minimum buffer concentration (the latter in order to facilitate product isolation).

Steroid conversions are often performed in the presence of solvents due to their low water solubility. Alcohols[58-60] or water-immiscible solvents[60-61] have been added to many gel-entrapped preparations (polyacrylamide, alginate, agar, photocross-linked resins, etc.). Omata et al.[62] immobilized thawed cells of *Nocardia rhodocrous* containing 3-beta-hydroxy steroid dehydrogenase in a number of photocross-linkable resin prepolymers or urethane prepolymers of varying hydrophobicity and tested the preparations on four different substrates in a benzene/*n*-heptane solvent mixture. The hydrophobic preparations showed significantly higher activities than the hydrophilic ones, and the effect was most pronounced for the steroids with an aliphatic side chain at position C-17 (most hydrophobic) since these were oxidized by hydrophobic gel preparations only. Recently, hydrophobic gels were also reported to be favorable for hydrolysis of dl-menthyl succinate in water-saturated *n*-heptane by entrapped *Rhodotorula minuta* cells[63] and for transglycosylation (production of adenine arabinoside, a slightly soluble antiviral agent) by entrapped *Enterobacter aerogenes.*[64] Furthermore, solvents have been added in order to solubilize indole in tryptophan production.[65] However, even when the processing parameters can be varied freely to optimum for biocatalyst utilization, the negative effects of microbial contamination will often urge the use of extreme pH values, extreme temperatures or the presence of solvents or preservatives to preclude growth. Naturally, the immobilization matrices must be resistant to such extreme conditions and preferably also to microbial attacks. Carbohydrate processing with glucose isomerase, inulinase or lactase (the latter when used in deproteinized whey) is preferentially carried out above 57°C, and the primary concern is to stabilize the cell preparations at these temperatures.

4. Cell Type and Enzyme Properties

When all the above-mentioned "nonenzyme" criteria have been allowed to limit the potential number of immobilization methods, the further selection can be attempted from optimized cell loading, activity yield, and operational stability plus, of course, economy.

The cell loading will naturally depend on the microorganism as well as the immobilization method. Messing et al.[66,67] have established the optimum pore size ranges of inorganic supports for maximum surface loading of biomass. They point out that the size distribution of a microorganism within the same culture may be wide (major dimension of *E. coli* was found to vary between 1 and 6 microns) and that additional space must be available if reproduction of the microorganisms within the carrier is desired. Such growth inside a carrier or gel matrix can be utilized for production of nonviable cell catalysts. Wada et al.[68] have demonstrated that growth inside K-carrageenan gel particles will produce a dense layer of cells close to the surface and result in a more active catalyst. Thus, the aspartase activity of an immobilized *E. coli* cell layer after cell rupture was reported to be 50% higher than that of the same number of homogeneously distributed immobilized cells.[68]

Table 3 compares cell loading by different immobilization methods. Adsorption or covalent binding to carriers will generally result in much lower loading (0 to 10 g wet cells/ℓ) than gel entrapment or cross-linking (100 to 1,000 g wet cells/ℓ), as measured from cell mass per catalyst volume (void volume excluded). Cross-linking and entrapment are therefore the methods of choice, in spite of a sometimes quite high diffusion restriction in gels and tightly cross-linked particles. Roels et al.[77] have even described how particle diffusion can be varied in gelatine-entrapped (glutaraldehyde cross-linked) cell preparations in order to tailor-make biocatalysts with predetermined half-life.

The location of the enzyme of interest within the cell and its intrinsic stability to various chemicals can have a profound influence on the activity yield. Cell surface-bound enzymes are more exposed to immobilization reagents (prior to cell rupture) and may require a mild immobilization procedure. Many of the published immobilization methods have been modified to allow such mild treatment. Thus, polymerization of acrylic monomers can be performed rapidly and be initiated by various chemicals[78,79] or by radiation,[80] ionotropic gels can be formed with different ions,[81] and glutaraldehyde cross-linking can be performed in the presence of inert proteins[41,82] or polyamines.[31] Addition of 0.4% polyethyleneimine to the whole broth of *Bacillus pasteurii* prior to cross-linking with glutaraldehyde raised the urease activity yield from 0.3% to 34% without impairing the physical stability of the preparation, and similar positive effects were found for other glutaraldehyde sensitive enzymes.[31]

Some of the mild immobilization methods may leave a high degree of residual cell integrity (or viability) in the preparations, which requires additional treatment to rupture the cells. Table 4 shows that cell rupture treatments can be performed before, during, or after immobilization, and that many positive effects have been achieved. Loss of viability *per se* is a desirable effect for many large-scale applications where growing cells would complicate the processing (the problems with microbial contaminations). Cell rupture before immobilization may lead to preparations with improved physical stabilities. Undesirable side activities, especially from cofactor-dependent enzymes or sensitive enzymes, may be removed by rupture, but can also be prevented by selection of microorganism strains or mutants lacking these activities (see Section IV.A.).

Table 3
CELL LOADING BY DIFFERENT IMMOBILIZATION METHODS

Immobilization method	Microorganism	Estimated cell loading (g wet cells/l wet catalyst)	Ref.
	Adsorption		
Fritted glass (3-6 μ pores)	*E. coli*	≃10	66
Ambersorb filter (3-23 μ pores)	*E. coli*	≃1	69
Dowex 1 resin	*E. coli*	<10	70
	Covalent Binding		
CM-cellulose + carbodiimide	*B. subtilis*	<1	24
Modified hydroxymethacrylate	*Zygosaccharomyces lactis*	1-2	71
	Entrapment		
Ca-alginate	*E. coli*	50-960[a]	72
K-carrageenan	*E. coli*	80	5, 68
Chitosan	*E. coli*	100-400[a]	57
Epoxy gel	*E. coli*	300-700[a]	72
Polyacrylamide, polymethacrylamide	*E. coli*	40-180	73, 55
Polyurethane foam	*E. coli*	20-100	74
Protein membrane	*E. coli*	≃40	75
Glutaraldehyde-tanned collagen	*C. simplex*	≃200	76
	Cross-Linking		
Polyethylene imine + glutaraldehyde	*B. subtilis*	500-1000[a]	31
Polyelectrolyte flocculation	*Arthrobacter sp.*	≤1000[a]	29

[a] Dried and rehydrated preparation.

Permeabilization or cell rupture is often a necessary condition for efficient utilization of the immobilized cell catalyst. The facilitated mass transport after rupturing treatment has been reported to give a 4 to 30 times increase in the expressed activities compared to intact cells. Yet, full activity is seldomly expressed by the immobilized nonviable cell preparations during application due to diffusion restrictions imposed by gel-entrapment or cross-linking.

For cofactor-dependent enzymes the most annoying negative effect of cell rupture is the loss of cofactors and cofactor regeneration, making it necessary to add cofactors continuously. Unfortunately native cofactors, especially NAD(H) and NADP(H), are prohibitively expensive for most large-scale applications, and not sufficiently stable to permit long-term re-use. Addition of electron acceptors like menadion (cheap) or phenazine methosulfate (expensive) has sometimes proved useful for steroid dehydrogenation reactions, but since the enzyme half-lives in the presence of synthetic cofactors have generally been short, the current tendency is to turn to viable cells for both steroid dehydrogenation[96,97] and hydroxylation.[37,88] On the other hand, addition of metal ions (e.g., Mg^{2+} for glucose isomerase[100] and aspartase[5]) and simple cofactors (e.g., pyridoxal phosphate for tryptophanase[65] and L-Asp-beta-decarboxylase[93]) has become an established practice with immobilized nonviable cells.

Table 4
CELL RUPTURE TREATMENTS

Treatment	Reported effect	Ref.
	Before Immobilization	
Drying	Removal of cofactor-dependent activities	83
	Cell clustering	71
	Increased sensitivity to immob. chemicals	78
Solvents	No need for activation of immob. cells	78
	Increased sensitivity to immob. chemicals	84
	Permeability to cofactors	59, 83
Freeze-thaw	Increased sensitivity to immob. chemicals	84
	Removal of cofactor-dependent activity	85
	No need for activation of immob. cells	86
Enzyme treatment (protease + cell wall degrading enzymes)	Cell clustering	71
Homogenization, sonication	Increased sensitivity to immob. reagents	84
	Improved physical stability of immobilized preparation	87
	During Immobilization	
Monomers and polymerization chemicals	No need for further permeabilization	23, 26,
	Removal of cofactor-dependent activities	78, 89
Cross-linking reagents	Removal of cofactor-dependent activities	88
	Increased permeability	88
	Increased operational stability	81
Solvents	No need for further permeabilization	90
	Loss of cofactor-dependent activities	58
	After Immobilization	
Drying	Increased physical stability	29, 87
	Facilitated mass transport	71
Solvents	Facilitated mass transport	91, 92
	Suppression of side activity	25, 81
Detergents	Suppression of side activity	25, 81
	Facilitated mass transport	24, 86
Cross-linking reagents	Increased mass transport, increased operational stability	93
Transition metals	Permeabilization to cofactors	71
Lysis, "activation" by incubation with substrate	Facilitated mass transport	86, 94, 95
Sonication/homogenization	Increased expressed activity	5

"Loss" of cofactors from permeabilized cells can also be utilized for cofactor production, and e.g., Murata et al.[98] have described how polyacrylamide-entrapped *S. cerevisiae*, but not intact cells, may continuously produce glutathione from its constituent amino acids and excrete it into the medium (Section V, Table 7).

5. Catalyst Manufacture

The final concern in selecting immobilization methods relates to the upscaling and flexibility of manufacture. Continuous or semicontinuous procedures may be preferred for larger volume catalysts, such as the manufacture of immobilized glucose isomerase and for immobilization of microorganisms produced by continuous fermentation. For the production of small volume catalysts a campaign manufacture may be chosen, and good storage stability of the catalyst becomes very important. Both wet and "dry" (e.g., maximum 10% water) storage of biocatalysts are presently used on the large

Table 5
COMPARISON OF MALIC ACID PRODUCTIVITY OF IMMOBILIZED *BREVIBACTERIUM AMMONIAGENES* AND *BREVIBACTERIUM FLAVUM*

Immobilization method and microbial cells	Fumarase activity of preparations (μmol h^{-1} g^{-1})	Operational stability at 37°C Half-life (days)	Relative productivity
Polyacrylamide			
B. ammoniagenes	5 800	53	100
B. flavum	6 680	94	204
Carrageenan			
B. ammoniagenes	5 800	75	142
B. flavum	9 920	160	516

From Takata, I., Yamamoto, K., Tosa, T., and Chibata, I., *Enzyme Microb. Technol.*, 2, 30, 1980. With permission.

scale, normally at refrigerator temperatures. While wet storage necessitates addition of preservatives, dry storage normally requires reswelling of the catalyst prior to loading into reactors, but has the advantage of cheaper shipping.

Catalyst manufacture costs will obviously depend on the number of unit operations involved, and consequently immobilization of cells is generally cheaper than immobilization of isolated intracellular enzymes. The matrices and immobilization reactants that have been preferred until now are all fairly cheap (U.S.$ 2-15 per kg), but some procedures such as acrylamide polymerization may require high safety equipment. When solvents are used, a solvent recovery operation will be added for economic and environmental reasons.

Chibata and co-workers have described the improvement of the aspartase catalyst by changing from *E. coli*/polyacrylamide to *B. flavum*/K-carrageenan and the numerous trials that were run before the new microorganism[20] and the new immobilization method[6] were selected. Without mentioning production or application security aspects or changed manufacturing costs, the Tanabe Seiyaku Company has reported a five times increased productitivity with the new biocatalyst which was taken into use in 1977. As seen from Table 5, the productivity gain primarily emanates from an increased operational stability as a synergistic effect of the new microorganism and the new immobilization method.

In conclusion, it should be evident that due to the large number of criteria that must influence the design of a new industrial biocatalyst there is never one single method or type of method that is outstanding and generally applicable for immobilization of nonviable cells.

D. Characterization of the Immobilized Cell Preparation

It is obviously almost impossible to compare characteristics between biocatalysts for different applications. Productivities of kg 6-APA/kg penicillin acylase at 37°C have nothing in common with productivities of kg (dry) high fructose corn syrup/kg glucose isomerase at 60°C. However, information on industrialized immobilized nonviable cell catalysts show that the expected productivities often lie in the range of 500 to 2,000 mol accumulated products per liter reactor volume and two half-lives (2 to 4 months).[6,9-13,17] In order to obtain a certain uniformity of basis information, a working party within the European Federation of Biotechnology has recently suggested a set of recommendations[99] including General Description of Biocatalyst, Method of Preparation, Physical and Chemical Characterization, and Kinetic Properties. Some of this information is found in the technical information sheets on commercial biocatalysts (see Ref. 9-12, 14, 17) since it is very important for the engineering design. Particle size and distribution, wet density, number of viable counts, volumetric activity under specified conditions, storage stability, and physical stability are often quality specifications, while production rates and operational stabilities will vary from one plant to another due to differences in substrate purity, cleaning procedures, process regulation, and number of shut-downs.

V. PRESENT AND FUTURE APPLICATIONS OF IMMOBILIZED NONVIABLE CELLS

The present industrial applications of immobilized nonviable cells are essentially covered by the commercial preparations listed in Table 1. For details on these applications reference is made to recent reviews on glucose isomerization,[100] lactose hydrolysis,[101] aspartate and malate production,[102] preparation and application of alpha-galactosidase and invertase,[103] and, finally, production of 6-APA and 7-ADCA[104,105] using immobilized biocatalysts.

Characteristical for today's commercial biocatalysts is that the enzymes have no requirements for complex (expensive) cofactors, and that they work on water-soluble substrates without presence of solvents or gases. The preparations are intended for use in fairly large reactors, and the biocatalyst cost lies in the range of 1 to 5% of the product sales prices.

Table 6 contains nearly 30 other enzymes that have been tested on the laboratory scale for utilization as immobilized nonviable cell preparations. A number of these enzymes are of the same "simple" type as the already commercialized enzymes, and inulinase, amino acid esterase, phosphatase, hydantoinase, His-ammonia lyase, aspartate decarboxylase, and urease may have a potential for upscaling in a situation with a sufficient market demand for a re-usable catalyst.[134]

In fact both hydantoinase (for D-*p*-hydroxy-phenylglycin production by Kanegafuchi) and aspartate decarboxylase (for L-alanine production by Tanabe Seiyaku Co.) are presently industrially used in the form of resting cells, and attempts are being made to develop immobilized catalysts. Aspartate decarboxylase (a pyridoxal phosphate-dependent enzyme) is especially interesting in combination with aspartase in order to convert ammonium fumarate directly to L-alanine, and the laboratory application of an immobilized two-cell system (*E. coli*/aspartase + *Ps. dacunhae*/aspartate decarboxylase) was recently described.[135]

Furthermore, resting cells are used in the production of L-Lysine where *Cryptococcus laurentii*/L-Amino caprolactamase and *Achromobacter obae*/Aminolactam racemase are added to the same vessel for conversion of DL-aminolactam to L-Lysine.[136] Also this two-cell system might be expected to benefit from immobilization.

Finally, *Kluyvera citrophila* (pen-G acylase) and *Pseudomonas melanogenum* (alpha-amino acid esterase/amidase) can be combined in order to produce ampicillin or amoxycillin from penicillin G in a two-step process. Kyowa Hakko has reported five to six times reuses of resting cells after recovery by centrifugation, but is now aiming at improving this by immobilization.[108]

Table 6 also contains a number of flavin-dependent oxidases (e.g., alcohol oxidase, glucose oxidase, xanthine oxidase) which are quite complex to use in the immobilized form due to their insufficient stability in the presence of H_2O_2 (this product is formed during re-oxidation of the flavin by oxygen). These enzymes require co-immobilization of catalase and/or another efficient peroxide degrading agent, and it has therefore been proposed to use oxidases in the form of cell preparations that naturally contain catalase (Table 7). Unfortunately catalases are also deactivated by peroxide (or super-

Table 6
UTILIZATION OF A SINGLE ENZYME IN IMMOBILIZED NONVIABLE CELLS

Enzyme	Microorganism/immobilization reagent or matrix	Ref.
Acetate kinase	*E. coli* B/polyacrylamide	106, 91
Alcohol oxidase	*Hansenula polymorpha*/polyacrylamide	86
Amino acylase	*Asp. ochraceus*/albumin, glutaraldehyde	107
Amino acid esterase	*Pseudomonas melanogenum*/not specified	108
Amino acid oxidase	*Trigonopsis variabilis* /Ca-alginate, MnO_2	109
Arginine deaminase	*Streptococcus faecalis*/polyacrylamide	85
	Pseudomonas putida/polyacrylamide	23
Aspartase	Cf. Table 1: *E. coli*/polyacrylamide	94
Aspartate decarboxylase	*Pseudomonas dacunhae*/K-carrageenan	93
Catalase	Yeast/polyacrylamide	84
Esterase	*Rhodotorula minuta*/photocross-linkable or urethane polymers	63
Fumarase	Cf. Table 1: *Brevibacterium ammoniagenes*/ polyacrylamide	25, 110
Alpha-galactosidase	Cf. Table 1 (Review 1980)	103
Beta-galactosidase	Cf. Table 1 (Reviews 1980, 1981)	101, 111
	Kluyveromyces lactis/polyethyleneoxide, glutaraldehyde	112
	Caldariella acidophila/cellulose acetate	113
	or albumin-glutaraldehyde	114
	or egg white-glutaraldehyde	115
Glucose isomerase	Cf. Table 1 (Reviews 1980)	116
Glucose oxidase	*Aspergillus niger*/not specified	117
Halohydrin epoxidase	*Flavobacterium sp.*/not specified	118
His-ammonia lyase	*Achromobacter liquidum*/polyacrylamide	26
	Micrococcus luteus/CMC-carbodiimide	24
	Pseudomonas fluorescens/hollow fiber	27
Hydantoinase	*Agrobacterium radiobacter*/not specified	119
Hydrogenase	*Proteus mirabilis* or *Clostridium sp.*/ gelatine, formaldehyde	120
	Alcaligenes eutrophus/Ca-alginate	121
Invertase	Cf. Table 1	
	Aspergillus or *Penicillium* spores/ ECTEOLA cellulose	35
	S. pasteurianus/agar pellets	90
	S. cerevisiae/gelatine	122
	Yeast/silica gel	123
Inulinase	*Kluyveromyces marxianus*/Ca-alginate	124
Beta-lactamase	*E. coli*/polyacrylamide	125
NAD kinase	*Brevibacterium ammoniagenes*/acrylate	126
NAD, NADP reductase	*Anacystis nidulans*/agar	127
	Alcaligenes eutrophus/Ca-alginate	121
Penicillin acylase	Cf. Table 1	
	E. coli/ionotropic polymers	56, 57, 68
	E. coli/silica gel	123
	E. coli/carrier cells	128
	E. coli/agar, glutaraldehyde	129
Phosphatase	*Aspergillus oryzae*/Celite	36
Polyphosphate glucokinase	*Achromobacter butyri*/polyacrylamide	91
Steroid dehydrogenases	*Arthrobacter simplex*/polyacrylamide	58
	Pseudomonas testosteroni/polyacrylamide	59
	Arthrobacter simplex/photocross-linkable resin polymer	60

Table 6 (continued)
UTILIZATION OF A SINGLE ENZYME IN IMMOBILIZED NONVIABLE CELLS

Enzyme	Microorganism/immobilization reagent or matrix	Ref.
	Nocardia rhodocrous/photocross-linkable resin prepolymers or urethane prepolymers	62
	Corynebacterium simplex/collagen, glutaraldehyde	75
Steroid oxidase	*Nocardia rhodocrous*/polyacrylamide or Ca-alginate	61
Tryptophanase	*E. coli*/polyacrylamide	65, 130
	E. coli/chitosan spheres	56
Beta-tyrosinase	*Erwinia herbicola*/collagen	131
Urease	*Bacillus pasteurii*/PEI, glutaraldehyde	31
	Bacillus pasteurii/sand, gelatine, glutaraldehyde	132
Xanthine oxidase	*Arthrobacter* X-4/gelatine, glutaraldehyde	133

oxide radicals), and this emphasizes the general problem with utilization of more than one enzyme in a cell preparation, i.e., that loss of one activity may render the whole system inactive. Brodelius et al.[109] combined *Trigonopsis variabilis* containing L-amino acid oxidase and catalase with MnO_2 in Ca-alginate beads and could then achieve an increased lifetime (days and weeks) for this catalyst for keto acid production. Thus, the present status of oxidase/catalase utilization in immobilized nonviable cells is that it may work on the laboratory scale, but that it remains to be proved that it can compete with other alternatives (e.g., soluble enzymes or conventional fermentations) on a larger scale.

The number of other immobilized nonviable cell preparations where more than one enzyme has been utilized is very limited, even on the laboratory scale (as seen from Table 7). However, the isomaltulose-forming catalyst[138] has been tested in pilot plant application and may be expected to become commercially utilized.

A novel group of interesting biocatalysts contains the cofactor converting enzymes. In the laboratory, NADP can now be produced from NAD both at the expense of ATP, utilizing only *Brevibacterium ammoniagenes*/NAD kinase[126] (Table 6), and at the expense of glucose plus initial ATP, utilizing the glycolysis pathway in yeast in combination with *Brevibacterium*/NAD kinase[143-144] (Table 7). Immobilized cells containing hydrogenases[121,127] have proved useful for reducing NAD and NADP in hydrogen atmosphere. ATP has been produced from adenosine or AMP by entrapped or encapsulated yeast,[139-141] and synthesis from precursors of other complex cofactors such as glutathione and CDP-choline is presently being studied by several Japanese groups with impressive results (Table 7). However, in spite of the high price level of the produced cofactors it is still uncertain if these immobilized nonviable cell preparations imply sufficient technical and economical benefits to compete with fermentative productions. One drawback is certainly the small market volume of the cofactors that make campaign batch production more favorable than continuous processing with a reusable catalyst.

Although it is many years since the "second generation" of immobilized biocatalysts was predicted to contain cofactor-dependent enzymes and cofactor regeneration,[145] the industrial breakthrough has not appeared, and cofactor-dependent enzymes are still exclusively used for analysis and laboratory synthesis. The new developments in industrial applications of immobilized nonviable cells during the next few years are also more likely to deal with nonaqueous solvents and two-enzyme utilization than with complex cofactor-dependent enzyme systems.

Table 7
UTILIZATION OF A FEW ENZYMES IN IMMOBILIZED NONVIABLE CELLS

Enzymes	Microorganism/immobilization reagent or matrix	Ref.
	Oxidase + Catalase	
Alcohol oxidase + catalase	*Hansenula polymorpha*/polyacrylamide	86
Amino acid oxidase + catalase	*Trigonopsis variabilis*/Ca-alginate	109
Cholesterol oxidase + catalase	*Nocardia rhodocrous*/polyacrylamide or Ca-alginate	61
Glucose oxidase + catalase	*Aspergillus niger*/not specified	117
Xanthine oxidase + catalase	*Arthrobacter* X-4/gelatine, glutaraldehyde	133
	Other Enzyme + Catalase	
Invertase + catalase (+ Co-immobilized isolated glucose oxidase)	Yeast/polyacrylamide	137
	Isomaltulose Forming Enzyme System	
Sucrose specific glucotransferase system (?)	*Erwinia rhapontici*/Ca-alginate	138
	Catabolic Sequence	
L-Arg. catabolic system	*Streptococcus faecalis*/polyacrylamide	85
	Synthesis Sequence	
CDP-choline synthesis	*Hansenula jadinii*/photocross-linked resin prepolymer	83, 139
ATP synthesis (from adenosine or AMP)	*S. cerevisiae*/polyacrylamide	140, 141
	S. cerevisiae/ethylcellulose	141
Glutathione synthesis	*S. cervisiae*/polyacrylamide	91, 141, 142
	Kinase + ATP Regenerating Enzyme(s)	
NAD kinase + glycolysis	*Brevibacterium ammoniagenes*/ cellulose acetate-butyrate + *S. cerevisiae*/cellulose acetatebutyrate	143
	Brevibacterium ammoniagenes/ polyacrylamide + *S. cerevisiae*/ polyacrylamide	144
	Synthetase(s) + ATP Regenerating Enzyme(s)	
CDP-choline synthetases + glycolysis	*S. cerevisiae*/ethylcellulose-chitosan	141
Glutathione synthetases + Acetate kinase	*E. coli* B/polyacrylamide	106
Glutathione synthetases + glycolysis	*S. cerevisiae*/polyacrylamide, *E. coli*/polyacrylamide + *S. cerevisiae*/polyacrylamide	144
CoA synthetases + glycolysis	*Brevibacterium ammoniagenes* + *S. cerevisiae*/polyacrylamide	141

REFERENCES

1. **Lilly, M. D.**, Production of intracellular microbial enzymes in *Applied Biochemistry and Bioengineering*, Vol. 2, Wingard, L. B., Katchalski-Katzir, E., and Goldstein, L., Eds., Academic Press, New York, 1979, chap. 1.
2. **Ruelius, H. W.**, Process for Producing an Enzyme System Capable of Degrading Penicillin G to 6-APA Using a Quarternary Ammonium Halide, U.S. Pat. 3,278,392, 1966.
3. **Singh, K., Sehgal, S. N., and Vézina, C.**, Hydrolysis of phenoxymethylpenicillin into 6-APA with fungal spores, *Appl. Microbiol.*, 17, 643, 1969.
4. **Kobayashi, H. and Suzuki, H.**, Kinetic studies of alpha-galactosidase containing mold pellets on PNPG hydrolysis, *Biotechnol. Bioeng.*, 18, 37, 1976.
5. **Sato, T., Nishida, Y., Tosa, T., and Chibata, I.**, Immobilization of *E. coli* cells containing aspartase activity with K-carrageenan, *Biochem. Biophys. Acta*, 570, 179, 1979.
6. **Takata, I., Yamamoto, K., Tosa, T., and Chibata, I.**, Immobilization of *Brevibacterium flavum* with carrageenan and its application for continuous production of L-malic acid, *Enzyme Microb. Technol.*, 2, 30, 1980.
7. Nippon Beet Sugar, Method of Immobilizing Endoenzyme, Jap. Pat. laid-open 64689, 1979.
8. **Stein, B. M. and Linden, J. C.**, Method of Stabilizing Alpha-galactosidase, US pat. 4,241,185, 1980.
9. NOVO Enzyme Information Novozym™ 231 — Immobilized Heatstable Lactase, IB 241b, 1981, NOVO Industri, DK-2880 Bagsvaerd, Denmark,
10. Technical Data Sheet, Maxazyme®, GI Immob., Gist-Brocades NV, Industrial Products Division, P.O. Box 1, 2600 MA Delft, Holland, 1979.
11. NOVO Enzyme Information, Sweetzyme®, B 174c, 1980, NOVO Industri, DK-2880 Bagsvaerd, Denmark.
12. Technical Information, Taka-Sweet™, Immobilized Glucose Isomerase for High Fructose Syrup Production, Miles Laboratories, Elkhart, Indiana, 46515 U.S.A., 1979.
13. **Takasaki, Y., Kosugi, Y., and Kanabayashi, A.**, *Fermentation Advances*, Academic Press, New York, 1969, 561.
14. Technical Data Sheet, Maxinvert™ Immob, Immobilized Invertase, Gist-Brocades NV, Industrial Products Division, P.O. Box 1, 2600 MA Delft, Netherlands, 1978.
15. **Chibata, I., Tosa, T., and Sato, T.**, 6-APA Microbial Production by Contacting Penicillins with Immobilized Penicillin Amidase Producing Microorganisms, Ger. Offen 2414 128, 1974.
16. **Nelson, R. P.**, Immobilized Microbial Cells, US Patent 3,957,580, 1976.
17. NOVO Enzyme Information, Novozym™ 217 — Immobilized Penicillin-V Acylase, IB 236b, 1981, NOVO Industri, DK-2880 Bagsvaerd, Denmark.
18. **Wang, Qing-Cheng, Ji, Xin-Song, and Yuan, Zhong-Yi**, Immobilization of Microbial Cells using Gelatine and Glutaraldehyde, Poster paper 1-11 presented at the 6th Enzyme Engineering Conf., Kashikojima, Japan, Sept. 1981.
19. **Aunstrup, K., Andresen, O., Falck, E. A., and Kjaer Nielsen, T.**, Production of microbial enzymes, in *Microbial Technology*, Vol. 1, Peppler, H. J. and Perlman, D., Eds., Academic Press, New York, 1979, chap. 9.
20. **Takata, I., Yamamoto, K., Tosa, T., and Chibata, I.**, Screening of microorganisms having high fumarase activity and their immobilization with carrageenan, *Eur. J. Appl. Microbiol. Biotechnol.*, 7, 161, 1979.
21. **Gestrelius, S.**, Immobilized penicillin-V acylase, development of an industrial catalyst, *Appl. Biochem. Biotech.*, 6, 1981.
22. General Standards for Enzyme Regulations, 1980, the Association of Microbial Food Enzyme Producers, Av. de Cortenbergh 172, 1040 Bruxelles.
23. **Yamamoto, K., Sato, T., Tosa, T., and Chibata, I.**, Continuous production of L-citrulline by immobilized *Pseudomonas putida* cells, *Biotechnol. Bioeng.*, 16, 1589, 1974.
24. **Jack, T. R. and Zajic, J. E.**, The enzymatic conversion of L-histidine to urocanic acid by whole cells of *Micrococcus luteus* immobilized on carbodiimide activated carboxymethylcellulose, *Biotechnol. Bioeng.*, 19, 631, 1977.
25. **Yamamoto, K., Tosa, T., Yamashita, K., and Chibata, I.**, Continuous production of L-malic acid by immobilized *Brevibacterium ammoniagenes* cells, *Eur. J. Appl. Microbiol. Biotechnol.*, 3, 169, 1976.
26. **Yamamoto, K., Sato, T., Tosa, T., and Chibata, I.**, Continuous production of urocanic acid by immobilized *Achromobacter liquidum* cells, *Biotechnol. Bioeng.*, 16, 1601, 1974.
27. **Kan, J. K. and Schuler, M. L.**, Urocanic acid production using whole cells immobilized in a hollow fiber reactor, *Biotechnol. Bioeng.*, 20, 217, 1978.
28. **Diers, I. and Emborg, C.**, Penicillin V-acylase, U.K. Pat. Appl. GB 2021119, 1979.
29. **Lee, C. K. and Long, M. E.**, Enzymatic Process Using Immobilized Microbial Cells, US Pat., Re 29,136, 1977.

30. **Kennedy, J. F.**, Facile methods for the immobilization of microbial cells without disruption of their life processes, in *Immobilized Microbial Cells*, Venkatsubramanian, K., Ed., ACS Symp. Ser. 106, Washington, 1979, chap. 8.
31. **Gestrelius, S.**, Immobilized Intracellular Enzymes, US Pat. 4,288,552, 1981.
32. **Gerson, D. F. and Zajic, J. E.**, The biophysics of cellular adhesion, in *Immobilized Microbial Cells*, Venkatsubramanian, K., Ed., ACS Symp. Ser. 106, Washington, 1979, chap. 3.
33. **Kennedy, J. F., Humphreys, J. D., Barker, S. A., and Greenshields, R. N.**, Application of living immobilized cells to the acceleration of the continuous conversions of ethanol to acetic acid — hydrous titanium (IV) oxide-immobilized *Acetobacter* species, *Enzyme Microb. Technol.*, 2, 209, 1980.
34. **Tso, Wung-Wai and Fung, Wai-Ping**, Bacterial organisms suitable for filamentous cell immobilization, *Biotechnol. Lett.*, 3, 421, 1981.
35. **Johnson, D. E. and Ciegler, A.**, Substrate conversion by fungal species entrapped in solid matrices, *Arch. Biochem. Biophys.*, 130, 384, 1969.
36. **Kuninaka, A., Rokugawa, K., and Yoshino, H.**, Conidia of *Aspergillus oryzae* as naturally immobilized phosphatases, *Agric. Biol. Chem.*, 44, 2825, 1980.
37. **Ohlson, S., Flygare, S., Larsson, P.-O., and Mosbach, K.**, Steroid hydroxylation using immobilized spores of *Curvularia lunata* germinated in situ, *Eur. J. Appl. Microbiol. Biotechnol.*, 10, 1, 1980.
38. **Krouwel, P. G., van der Laan, W. F. M., and Kossen, N. W. F.**, A method for starting a fermenter with immobilized spores of the genus *Clostridium* aseptically, *Biotechnol. Lett.*, 3, 158, 1981.
39. **Hirano, K.-I., Karube, I., and Suzuki, S.**, Amino Acylase pellets, *Biotechnol. Bioeng.*, 19, 311, 1979.
40. EEC Council Directive of 17 September 1979 on the sixth amendment of the Directive 67/548/EEC on the approximation of laws, regulations and administrative provisions relating to the classification, packaging and labelling of dangerous substances.
41. **Vieth, W. R. and Venkatsubramanian, K.**, Collagen-immobilized enzyme systems, in *Methods in Enzymology*, Vol. 44, Mosbach, K., Ed., Academic Press, New York, 1976, chap. 19.
42. **Bartoli, F., Morisi, F., and Zaccardelli, D.**, Radial Reactor for Carrying Out Enzyme-Catalyzed Reactions, German Offen Pat. 2844688, 1979.
43. **Hamsher, J. J. and Lozanov, M.**, Process for preparing 6-APA, US pat. 4,113,566, 1978.
44. **Lilly, M. D.**, Immobilized enzyme reactors, *Biotechnology*, 1693, 82, 165, 1978.
45. **Norsker, O., Gibson, K., and Zittan, L.**, Experience with empirical methods for evaluating pressure drop properties of immobilized glucose isomerase, *Starch/Starke*, 31, 13, 1979.
46. **Cheetham, P., Blunt, K., and Bucke, C.**, Physical studies on cell immobilization using calcium alginate gels, *Biotechnol. Bioeng.*, 21, 2155, 1979.
47. **Klein, J. and Kluge, B.**, Packed bed flow resistance, in *Characterization of Immobilized Biocatalysts*, Vol. 84, Buchholz, K., Ed., Dechema Monographs No. 1724-1731, Weinheim, 1979, chap. 6.3.2.
48. **Klein, J. and Washausen, P.**, Pressure stability (compression experiment), in *Characterization of Immobilized Biocatalysts*, Buchholz, K., Ed., Weinheim, 1979, chap. 6.3.1.
49. **Cheetham, P.**, Physical studies on the mechanical stability of columns of calcium alginate gel pellets containing entrapped microbial cells, *Enzyme Microb. Technol.*, 1, 183, 1979.
50. **Klein, J. and Eng, H.**, The measurement of the abrasion, in *Characterization of Immobilized Biocatalysts*, Vol. 84, Buchholz, K., Ed., Dechema Monographs No. 1724-1731, Weinheim, 1979, chap. 6.3.3.
51. **Chen, A. H. and Jao, Y.-C.**, Miles Laboratories Inc., Preparation of a Bacterial Cell Aggregate, US patent 4,241,632, 1981.
52. **Borglum, G. B.**, Miles Laboratories Inc., Production of Bacterial Cell Aggregate, U.K. Pat. Appl. GB 2,033,396, 1980.
53. **Birnbaum, S., Larsson, P.-O., and Mosbach, K.**, Stabilization of Calcium Alginate Gel, Poster Presented at the 2nd European Congress of Biotechnology, Eastbourne, April 1981.
54. **Veliky, I. A. and Williams, R. E.**, The production of ethanol by *Saccharomyces cerevisiae* immobilized in polycation-stabilized calcium alginate gels, *Biotechnol. Lett.*, 3, 275, 1981.
55. **Takata, I., Tosa, T., and Chibata, I.**, Screening of matrix suitable for immobilization of microbial cells, *J. Solid Phase Biochem.*, 2, 225, 1977.
56. **Klein, J. and Wagner, F.**, Immobilized whole cells, in *Biotechnology*, Proc. 1st Eur. Congress on Biotechnol., Dechema Monographs No. 1693-1703, Vol. 82, Weinheim, 1978, 142.
57. **Vorlop, K.-D. and Klein, J.**, Formation of spherical chitosan biocatalysts by ionotropic gelation, *Biotechnol. Lett.*, 3, 9, 1981.
58. **Larsson, P.-O. and Mosbach, K.**, Immobilization of steroid-transforming microorganisms in polyacrylamide, in *Methods in Enzymology*, Vol. 44, Mosbach, K., Ed., Academic Press, New York, 1976, chap. 13.
59. **Yang, H. S. and Studebaker, J. F.**, Continuous dehydrogenation of a steroid with immobilized microbial cells: effect of an exogeneous electron acceptor, *Biotechnol. Bioeng.*, 20, 17, 1978.
60. **Sonomoto, K., Tanaka, A., Omata, T., Yamane, T., and Fukui, S.**, Application of photo-crosslinkable prepolymers to entrap microbial cells, *Eur. J. Appl. Microbiol. Biotechnol.*, 6, 325, 1979.

61. **Duarte, J. M. C. and Lilly, M. D.**, The use of free and immobilized cells in the presence of organic solvents, in *Enzyme Engineering 5*, Weetall, H. H. and Royer, G. P., Eds., Plenum Press, New York, 1980, 363.
62. **Omata, T., Tida, T., Tanaka, A., and Fukui, S.**, Transformation of steroids by gel-entrapped *Nocardia rhodocrous* Cells in organic solvent, *Eur. J. Appl. Microbiol. Biotechnol.*, 8, 143, 1979.
63. **Omata, T., Iwamoto, N., Kimura, T., Tanaka, A., and Fukui, S.**, Stereoselective hydrolysis of dl-menthyl succinate by gel-entrapped *Rhodotorula minuta* var. *Texensis* cells in organic solvent, *Eur. J. Appl. Microbiol. Biotechnol.*, 11, 199, 1981.
64. **Fukui, S. and Tanaka, A.**, Bio-Conversion of Lipophilic or Water-Insoluble Compounds by Immobilized Biocatalysts in Organic Solvents, paper presented at 6th Enzyme Engineering Conf., Kashikojima, Japan, Sept. 1981.
65. **Decottignies-Le Maréchal, P., Caldéron-Seguin, R., Vandecasteele, J. P., and Azerad, R.**, Synthesis of L-tryptophan by immobilized *E. coli* cells, *Eur. J. Appl. Microbiol. Biotechnol.*, 7, 33, 1979.
66. **Messing, R. A. and Oppermann, R. A.**, Pore dimension for accumulating biomass. I. Microbes that reproduce by fission or by budding, *Biotechnol. Bioeng.*, 21, 49, 1979.
67. **Messing, R. A., Oppermann, R. A., and Kolot, F. B.**, Pore dimensions for accumulating biomass. II. Microbes that form spores and exhibit mycelial growth, *Biotechnol. Bioeng.*, 21, 59, 1979.
68. **Wada, M., Kato, J., and Chibata, I.**, A new immobilization of microbial cells, *Eur. J. Appl. Microbiol. Biotechnol.*, 8, 241, 1979.
69. Rohm and Haas Company (Philadelphia, Penn. 19105, U.S.A.), Preliminary Technical Notes Ambersorb™ XE-342 and Ambersorb™ XE-352 Filters, April 1977.
70. **Hattori, T. and Furusaka, C.**, Chemical activities of *E. coli* adsorbed on a resin, *J. Biochem. (Japan)*, 48, 831, 1960.
71. **Jirku, V., Turkova, J., Kuchynkova, A., and Krumphanzl, V.**, Modified hydroxyalkylmethacrylate gel as a support for immobilization of yeast cells, *Eur. J. Appl. Microbiol. Biotechnol.*, 6, 217, 1979.
72. **Klein, J., Washausen, P., Kluge, M., and Eng, H.**, Physical characterization of biocatalyst particles obtained from polymer entrapment of whole cells, in *Enzyme Engineering*, Vol. 5, Weetall, H. H. and Royer, G. P., Eds., Plenum Press, New York, 1980, 359.
73. **Sato, T., Mori, T., Tosa, T., Chibata, I., Furui, M., Yamshita, K., and Sumi, A.**, Engineering analysis of continuous production of L-aspartic acid by immobilized *E. coli* cells in fixed beds, *Biotechnol. Bioeng.*, 17, 1797, 1975.
74. **Klein, J. and Kluge, M.**, Immobilization of microbial cells in polyurethane matrices, *Biotechnol. Lett.*, 3, 65, 1980.
75. **Petre, D., Noel, C., and Thomas, D.**, A new method for cell immobilization, *Biotechnol. Bioeng.*, 20, 127, 1978.
76. **Constantinides, A.**, Steroid transformation of high substrate concentrations using immobilized *Corynebacterium simplex* cells, *Biotechnol. Bioeng.*, 22, 119, 1980.
77. **Roels, J. A. and van Tilburg, R.**, Gist-Brocades, Operational Stability and Productivity of an Immobilized Glucose Isomerase, lecture presented at VIth Fermentation Symp., London, Ont., Canada, July, 1980.
78. **Martin, C. K. A. and Perlman, D.**, Conversion of L-sorbose to L-sorbosone by immobilized cells of *Gluconobacter melanogenes* IFO 3293, *Biotechnol. Bioeng.*, 18, 217, 1976.
79. **Chibata, I., Tosa, T., and Sato, T.**, Immobilized aspartase-containing microbial cells: preparation and enzymatic properties, *Appl. Microbiol.*, 27, 878, 1974.
80. **Kawashima, K. and Umeda, K.**, Preparation of membranous immobilized invertase and its characteristics, *Agr. Biol. Chem.*, 40, 1151, 1976.
81. **Tosa, T., Sato, T., Mori, T., Yamamoto, K., Takata, I., Nishida, Y., and Chibata, I.**, Immobilization of enzymes and microbial cells using carrageenan as matrix, *Biotechnol. Bioeng.*, 21, 1697, 1979.
82. **van Velzen, A. G.**, Water-insoluble Enzyme Composition, U.S. Pat. 3,838,007, 1974.
83. **Fukui, S., Tanaka, A., and Gellf, G.**, Immobilization of enzymes, microbial cells, and organelles by inclusion with photocrosslinkable resins, in *Enzyme Engineering 4*, Broun, G. B., Manecke, G., and Wingard, Jr., L. B., Eds., Plenum Press, New York, 1978, 299.
84. **D'Souza, S. F. and Nadkarni, G. B.**, Immobilized catalase-containing yeast cells: preparation and enzymatic properties, *Biotechnol. Bioeng.*, 22, 2191, 1980.
85. **Franks, N. E.**, Catabolism of L-Arg by polyacrylamide entrapped *Streptococcus faecalis* ATCC 8043, *Biotechnol. Bioeng. Symp.*, 3, 327, 1972.
86. **Couderc, R. and Baratti, J.**, Immobilized yeast cells with methanol oxidase activity: preparation and enzymatic properties, *Biotechnol. Bioeng.*, 22, 1155, 1980.
87. **Amotz, S., Kjaer Nielsen, T., Thiesen, N. O.**, Immobilization of Glucose Isomerase, US Pat. 3,980,521, 1976.
88. **De Rosa, M., Gambacorta, A., Lama, L., Nicolaus, B., and Buonocore, V.**, Immobilization of thermophilic microbial cells in crude egg white, *Biotechnol. Lett.*, 3, 183, 1981.

89. **Maddox, I. S., Dunnill, P., and Lilly, M. D.,** Use of immobilized cells of *Rhizopus nigricans* for the 11-alpha-hydroxylation of progesterone, *Biotechnol. Bioeng.*, 23, 345, 1981.
90. **Toda, K. and Shoda, M.,** Sucrose inversion by immobilized yeast cells in a complete mixing reactor, *Biotechnol. Bioeng.*, 17, 481, 1975.
91. **Murata, K., Tani, K., Kato, J., and Chibata, I.,** Continuous production of glutathione using immobilized microbial cells containing ATP generating systems, *Biochimie*, 62, 347, 1980.
92. **Murata, K., Uchida, T., Tani, K., Kato, J., and Chibata, I.,** Continuous production of glucose-6-phosphate by immobilized *Achromobacter butyri* cells, *Eur. J. Appl. Microbiol. Biotechnol.*, 7, 45, 1979.
93. **Yamamoto, K., Tosa, T., and Chibata, I.,** Continuous productions of L-alanine using *Pseudomonas dacunhae* immobilized with carrageenan, *Biotechnol. Bioeng.*, 22, 2045, 1980.
94. **Chibata, I., Tosa, T., and Sato, T.,** Immobilized aspartase-containing microbial cells: preparation and enzymatic properties, *Appl. Microbiol.*, 5, 878, 1974.
95. **Vieth, W. R., Venkatsubramanian, K., and Constantinides, A.,** Continuous Synthesis of Organic Acids and Transformation of Steroids by Immobilized Microbial Cells, Abstracts 15.26, 5th Int. Fermentation Symp., Berlin, 1976.
96. **Koshcheenko, K. A., Sukhodolskaya, G. V., Tyurin, V. S., and Skryabin, G. K.,** Physiological, biochemical and morphological changes in immobilized cells during repeated periodical hydrocortisone transformations, *Eur. J. Appl. Microbiol. Biotechnol.*, 12, 161, 1981.
97. **Sawada, H., Kinoshita, S., Yoshida, T., and Taguchi, H.,** Continuous production of 12-ketochenodeoxycholic acid in a column reactor containing immobilized living cells of *Brevibacterium fuscum*, *J. Ferment. Technol.*, 59, 111, 1981.
98. **Murata, K., Tani, K., Kato, J., and Chibata, I.,** Glutathione production by immobilized *Saccharomyces cerevisiae* cells containing an ATP regeneration system, *Eur. J. Appl. Microbiol. Biotechnol.*, 11, 72, 1981.
99. European Federation of Biotechnology, Working Party on Immobilized Biocatalysts, Proposal for a General Characterization of Immobilized Biocatalysts, poster Presented at the 2nd European Congress of Biotechnology, Eastbourne, April, 1981.
100. **MacAllister, R. V.,** Manufacture of high fructose corn syrup using immobilized glucose isomerase, in *Immobilized Enzymes for Food Processing*, Pitcher, W.-H., Jr., Ed., CRC Press, Boca Raton, Fla., 1980, chap. 4.
101. **Coughlin, R. W. and Charles, M.,** Applications of lactase and immobilized lactase, in *Immobilized Enzymes for Food Processing*, Pitcher, W.-H., Jr., Ed., CRC Press, Boca Raton, Fla., 1980, chap. 6.
102. **Sharma, B. P. and Messing, R. A.,** Application and potential of other enzymes in food processing: aminoacylase, aspartase, fumarase, glucose oxidase-catalase, in *Immobilized Enzymes for Food Processing*, Pitcher, W.-H., Jr., Ed., CRC Press, Boca Raton, Fla., 1980, chap. 8.
103. **Reilly, P. J.,** Potential and use of immobilized carbohydrases, in *Immobilized Enzymes for Food Processing*, Pitcher, W.-H., Jr., Ed., CRC Press, Boca Raton, Fla., 1980, ch. 5.
104. **Johnson, J. C.,** *Industrial Enzymes Recent Advances*, Noyes Data Corp., Park Ridge, N.J., 1977, 287.
105. **Moss, M. O.,** Enzymic alterations of penicillins and cephalosporins, in *Topics in Enzyme and Fermentation Technology*, Wiseman, A., Ed., Ellis Horwood Ltd., Chichester, 1977, chap. 5.
106. **Murata, K., Tani, K., Kato, J., and Chibata, I.,** Application of Immobilized ATP in the production of glutathione by a multienzyme system, *J. Appl. Biochem.*, 1, 283, 1979.
107. **Hirano, K.-I., Karube, I., and Suzuki, S.,** Amino acylase pellets, *Biotechnol. Bioeng.*, 19, 311, 1977.
108. **Okachi, R., Hashimoto, Y., Kawamori, M., Katsumata, R., Takayama, K., and Nara, T.,** Enzymatic Synthesis of Penicillins and Cephalosporins by Penicillin Acylase, paper presented at the 6th Enzyme Engineering Conf., Kashikojima, Japan, Sept. 1981.
109. **Brodelius, P., Hägerdal, B., and Mosbach, K.,** Immobilized whole cells of the yeast *Trigonopsis variabilis* containing D-amino acid oxidase for the production of alpha-keto acids, in *Enzyme Engineering 5*, Weetall, H. H. and Royer, G. P., Eds., Plenum Press, New York, 1980, 383.
110. **Yamamoto, K., Tosa, T., Yamashita, K., and Chibata, I.,** Kinetics and decay of fumarase activity of immobilized *Brevibacterium ammoniagenes* cells for continuous production of L-malic acid, *Biotechnol. Bioeng.*, 19, 1101, 1977.
111. **Greenberg, N. A. and Mahoney, R. R.,** Immobilization of lactase for use in dairy processing: a review, *Proc. Biochem.*, Feb./March, 2, 1981.
112. **Jirku, V., Turkova, J., Veruovic, B., and Kubánek, V.,** Immobilization of yeast cells on polyphenylene oxide, *Biotechnol. Lett.*, 2, 451, 1980.
113. **De Rosa, M., Gambacorta, A., Esposito, E., Drioli, E., and Gaeta, S.,** Thermophilic microbial cells immobilized in cellulose acetate membranes, *Biochimie*, 62, 517, 1980.
114. **Drioli, E., Iorio, G., and Molinari, R.,** High-temperature membrane-entrapped cells, *Biotechnol. Bioeng.*, 23, 221, 1981.

115. De Rosa, M., Gambacorta, A., Lama, L., Nicolaus, B., and Buonocore, V., Immobilization of thermophilic microbial cells in crude egg white, *Biotechnol. Lett.*, 3, 183, 1981.
116. Chen, W.-P., Glucose isomerase (a review), *Proc. Biochem.*, Aug./Sept., 36, 1980.
117. Richter, G. and Heinecker, H., Conversion of Glucose to Gluconic Acid by Means of Immobilized Glucose Oxidase, paper presented at Detmold Starch Convention, 1979.
118. Neidleman, S. L., A Unique Multi-Enzyme Process for the Production of Commodity Chemicals, paper presented at the Symp. on Biochemical Approaches to Organic Synthesis, Jerusalem, August, 1980.
119. Olivieri, R., Fascetti, E., Angelini, L., and Degen, L., Microbial transformation of racemic hydantoins to D-amino acids, *Biotechnol. Bioeng.*, 23, 1981, in press.
120. Tischer, W., Tiemayer, W., and Simon, H., Stereospecific hydrogenations with immobilized microbial cells or enzymes, *Biochimie*, 62, 331, 1980.
121. Klibanov, A. M. and Puglisi, A. V., The regeneration of coenzymes using immobilized hydrogenase, *Biotechnol. Lett.*, 10, 445, 1980.
122. Parascandola, P. and Scardi, V., Gelatin-entrapped whole-cell invertase, *Biotechnol. Lett.*, 3, 369, 1981.
123. Tschang, C.-J., Klofenz, H., Sanner, A., and Zahn, W., Production of Silica Gel Entrapped Enzyme Preparations, Ger. Offen 2911776, 1980.
124. Kierstan, M. and Buche, C., The immobilization of microbial cells, subcellular organelles, and enzymes in calcium alginate gels, *Biotech. Bioeng.*, 19, 387, 1977.
125. Pache, W., Degradation of beta-lactam antibiotics by polyacrylamide-entrapped beta-lactamase producing *E. coli* cells, *Eur. J. Appl. Microbiol.*, 5, 171, 1978.
126. Hayashi, T., Tanaka, Y., and Kawashima, K., Immobilization of microbial cells containing NAD-kinase, *Biotechnol. Bioeng.*, 21, 1019, 1979.
127. Weetall, H. H. and Krampitz, L. O., Production of hydrogen from water using biophotolytic methods, *J. Solid-Phase Biochem.*, 2, 115, 1980.
128. Vojtisek, V., Zeman, R., Barta, M., Culik, K., Chaloupka, J., Katal, J., Drobnik, J., and Svec, F., Microbial Cell Aggregates, Ger. Offen 2833071, 1979.
129. Wang Zhen-xiang, Yen Hua-ai, Wang Mei-zhi, Jiao Quing-hua, Han Wen-zhen, Sun Wan-ru, and Zhang Qi-xian, Production of 7-ADCA by Immobilized *E. coli* Cells, poster paper 1-26 presented at the 6th Enzyme Engineering Conf., Kashikojima, Japan, Sept. 1981.
130. Azerad, R., Calderón-Seguin, R., and Decottignies-Le Maréchal, P., Production de L-Tryptophane par des Bactéries Immobilisées, *Bull. Soc. Chim.*, Fr. No. 1-2, 83, 1980.
131. Yamada, H., Yamada, K., Kumagai, H., Hino, T., and Okamura, S., Immobilization of beta-tyrosinase cells with collagen, in *Enzyme Engineering*, Vol. 3, Pye, E. K. and Weetall, H. H., Eds., Plenum Press, New York, 1978, 57.
132. Branner-Jørgensen, S., Enzyme Immobilization, U.S. Pat. 4,266,029, 1981.
133. Tramper, J., Van der Plas, H. C., Van der Kaaden, A., Müller, F., and Middelhoven, W. J., Xanthine oxidase activity of *Arthrobacter* X-4 cells immobilized in glutaraldehyde-cross-linked gelatin, *Biotechnol. Lett.*, 1, 397, 1979.
134. Poulsen, P. B., European and American trends in the industrial application of immobilized biocatalysts, *Enzyme Microb. Technol.*, 3, 271, 1981.
135. Sato, T., Takamatsu, S., Yamamoto, K., Umemura, I., Tosa, T., and Chibata, I., Production of L-Alanine from Ammonium Fumarate Using Two Immobilized Microbial Cells, poster paper 1-25 presented at the 6th Enzyme Engineering Conf., Kashikojima, Japan, Sept. 1981.
136. Fukumura, T., Conversion of D- and DL-alpha-Amino-epsilon-caprolactam into L-Lysine using both yeast cells and bacterial cells, *Agric. Biol. Chem.*, 41, 1327, 1977.
137. D'Souza, S. F. and Nadkarni, G. B., Continuous conversion of sucrose to fructose and gluconic acid by immobilized yeast cell multienzyme complex, *Biotechnol. Bioeng.*, 22, 2179, 1980.
138. Bucke, C. and Cheetham, P., Tate and Lyle Ltd., Isomaltulose Production from Sucrose, European Patent Appl. 28900, 1981.
139. Kimura, A., Tanaka, A., Matsuno, R., and Fukuda, H., Immobilization of glycolysis system of yeasts and production of cytidine dipohosphate choline, *Eur. J. Appl. Microbiol. Biotechnol.*, 5, 13, 1978.
140. Asada, M., Morimoto, K., Nakanishi, K., Matsuno, R., Tanaka, A, Kimura, A., and Kamikubo, T., Continuous ATP regeneration using immobilized yeast cells, *Agric. Biol. Chem.*, 43, 1773, 1979.
141. Samejima, H., Kimura, K., Ado, Y., Suzuki, Y., and Tadokoro, T., Regeneration of ATP by immobilized microbial cells and its utilization for the synthesis of nucleotides, in *Enzyme Engineering*, Vol. 4, Broun, G. B., Manecke, G., and Wingard, L. B., Eds., Plenum Press, New York, 1978, 237.
142. Murata, K., Tani, K., Kato, J., and Chibata, I., Continuous production of glutathione by immobilized *Saccharomyces cerevisiae* Cells, *Eur. J. Appl. Microb. Biotech.*, 6, 23, 1978.

143. **Ado, Y., Kimura, K., and Samejima, H.**, Production of useful nucleotides with immobilized microbial cells, in *Enzyme Engineering*, Vol. 5, Weetall, H. H. and Royer, G. P., Eds., Plenum Press, New York, 1980, 295.
144. **Murata, K., Tani, K., Kato, J., and Chibata, I.**, Glycolytic pathway as an ATP generation system and its application to the production of glutathione and NADP, *Enz. Microb. Technol.*, July, 3, 1981.
145. **Wingard, L. B.**, Developments and challenge of enzyme engineering, in *Enzyme Engineering*, Vol. 2, Pye, E. K. and Wingard, L. B., Eds., Plenum Press, New York, 1974, 3.

Chapter 2

IMMOBILIZED VIABLE CELLS

Bo Mattiasson

TABLE OF CONTENTS

I. VIABLE IMMOBILIZED CELLS — A DEFINITION

It may be difficult to find a clear and simple definition for the viability of immobilized cells. In many cases it has been reported that cells metabolize glucose in the presence of oxygen to produce carbon dioxide and water. This ability is, in my mind, only an indication that a major part of the main metabolic routes must still be operational and furthermore that the membrane-function must be fairly good since coenzymes are still available for the enzymes involved in these metabolic routes. Whether or not the cells are viable still remains a question.

Another criterion that has been used to estimate cellular viability has been that of protein synthesis, which can be expressed, for example, as an ability to generate new enzymes when the activity of the cell is decreasing. In this case a major part of cellular metabolism must also be intact, but is it enough to judge whether or not the cell is viable?

One alternative for determining viability is to estimate whether or not the cell mass in the immobilized preparation increases. This can be done by dry weight measurements or by protein quantification[1] on the immobilized cell preparation. Both these measurements give indirect information of cellular viability.

In other cases researchers have used growth on nutrient agar plates to indicate viability. Cells capable of forming new colonies have been recognized as being alive. However, even this apparently straightforward definition has some exceptions. Thus, when measuring the viability of yeast cells recovered from a preparation trapped in polyacrylamide, it turned out that after an incubation of the standardized time period only very few colonies were seen, whereas after an extended period of incubation much higher cell counts were obtained. Obviously a fraction of the cells were "dormant" or poisoned so that a longer time was needed for cell growth to take place.[2] It is realistic to assume that in many cases cells are in a condition far from what is normally expected, but they still have the ability to recover.

As seen from the above example, it may be difficult to get a clear-cut definition of viability, and this may be the reason why in many cases, it has been difficult to judge, from published data, if the cells are alive or not. Table 1 lists some of the criteria used to define viability of immobilized cells, and references where they have been applied.

II. WHY LIVING CELLS?

The objective behind immobilizing biologically active species is clear today. Reactions based on immobilized cell preparations rather than conventional fermentation technology offer advantages such as:

1. Higher reaction rates due to increased cell densities
2. Higher specific product yield
3. Possibilities for continuous operation
4. High dilution rates without the problem of washout
5. Better and easier control
6. Reduced demands on costly fermentors
7. Easier product up-grading

With regard to single-step enzyme reactions, it turned out to be advantageous in some cases to use whole cells instead of purified enzymes. Besides avoiding the troublesome and expensive purification of the enzymes in question, these enzymes could be kept in their natural, intracellular milieu. It has, however, turned out that in many of the cases

Table 1
CRITERIA USED FOR DEFINING VIABILITY OF IMMOBILIZED CELLS

Criteria	Organism	Immobilization methods used	Ref.
Microscopic observation	*Saccharomyces cerevisiae*	Entrapment in Ca-alginate	3
Substrate consumption	*Candida tropicalis*	Adsorption to a ceramic support	4
Oxygen uptake	*Escherichia coli* *Saccharomyces cerevisiae*	Partial covalent binding to hydrous zirkonium(IV) and titanium(IV) oxides	5
Measurement of nitrogen content	*Candida tropicalis*	Entrapment in copoly-(styrene-nucleic acid)/Al^{3+} network	6
Reincubation results in increased activity	*Curvularia unata*	Entrapment in polyacrylamide	7
	Pseudomonas putida	Entrapment in polyacrylamide	8
Increase in cell numbers			
Gravimetric	*Arthrobacter simplex*	Entrapment in polyacrylamide	9
Microscopic observation	*Bacillus subtilis*	Entrapment in polyacrylamide	10
Counting of cells after liberation from the gel matrix	*Saccharomyces cerevisiae*	Entrapment in polyacrylamide	2
Counting of cells after liberation from gel matrix	*Saccharomyces cerevisiae*	Entrapment in Ca-alginate	3
	Penicillium chrysogenum	Entrapment in polyacrylamide	11
Response of the reactivation process to antibiotics	*Arthrobacter simplex*	Entrapment in polyacrylamide	9

the cells have lysed after immobilization, leaving the enzyme entrapped in a cavity with a mixture of other microbial proteins.[12] Still, increased enzymatic stabilities have been reported, in these cases compared with immobilized, purified enzymes. The advantages for using viable cells were perhaps not clear initially, since at that time the objective was to use as mild immobilization conditions as possible in order to preserve the activity of even the most labile enzymes. It then turned out that cell division was possible upon administration of substrate to some of the resulting immobilized cell preparations. Viable cells may thus function by maintaining the catalytic activities of labile enzymes, e.g., induced enzymes, and furthermore, they may be of value when sequential reactions involving several enzymes as well as coenzymes are studied. In this context it must be stressed that the viable cells possess the capacity to regenerate and retain the coenzymes which are required for some catalytic reactions. Immobilization of cells is the only general method available today for solving two of the main problems in enzyme technology, namely coenzyme regeneration and retention, even if other technologies are currently under development (see Chapter 1, Volume II).

As indicated above, cells that are losing their activity may be "recharged" by a short period of incubation with growth medium. This makes it possible to increase the operational lifetime for a certain preparation. Furthermore, it is also possible to induce enzyme activity in the immobilized cells.

On the negative side, the increase in cell mass which occurs when cells proliferate will cause problems by allowing cell leakage into the medium. Furthermore, when entrapped cells are used, the interior of the support will be crowded by cells leading to an ultimate breakage of the support from the inside in extreme cases.

III. PHYSIOLOGY OF IMMOBILIZED CELLS

As a general rule it can be stated that biocatalysts undergo a loss of expressed activity upon immobilization; their effectiveness factor decreases. This has been ascribed to diffusional restrictions. In the case of immobilized enzymes, much work has been put into developing an understanding of the effects on the catalytic efficiency of enzymes exerted by their microenvironments, as well as the effects on the microenvironment by the catalytic process. From such studies has emerged an insight in how it is possible to manipulate the environment in order to influence the enzyme activity in any special direction.[13,14]

Much of the experience obtained from enzymes can, of course, be directly applied to the situation with immobilized cells, but here some complications arise. The cells *per se* generate a microenvironment for the individual enzyme activities. The intact cell membrane protects these enzymes from drastic changes in environmental parameters. However, in the case of viable cells, changes in the environment will influence the entire metabolism, not only the enzyme(s) of specific interest for a certain application.

Physical as well as chemical influences will thus cause changes in the total metabolism of the immobilized cells. So far reports usually state observations of an unpredicted metabolic behavior of the immobilized cells but with no attempts to interpret the situation.

Another basic problem arises when high densities of aerobic cells are used, and that is one of oxygen transfer. The fact that immobilized cells often are poorly supplied with oxygen has led to a preference for anaerobic cells when designing research projects. When aerobic cells have been used, rather low cell densities have been a necessity in order to ensure that oxygen supply is adequate. However, it is realistic to assume that immobilized aerobic cells used in may cases have operated under oxygen deficiency (see also Chapter 3, this volume).

A recent paper[15] reported on α-amylase production and on the physiology of immobilized *Bacillus amyloliquefacieus.* The bacteria were immobilized in K-carrageenan (34 g/ℓ of polymer) in bead form and the production of α-amylase was studied.

It was found that the respiratory activity of the entrapped cells (3 mg cell dry weight/g wet gel) was repressed by about 50% as compared to that registered after releasing the cells from the carrageenan beads. In spite of the repression, the oxygen uptake rate per unit volume of gel beads was constant, suggesting that the diffusion of oxygen into the gel might be the limiting step in the respiration of entrapped cells. Figure 1 schematically illustrates the main metabolic routes of an aerobic cell, and also gives some indications for possible metabolic changes due to a decreased oxygen supply.

In their studies on *Bacillus amyloliquefacieus,* Shinmyo et al.[15] found that cell growth of entrapped cells was also repressed, whereas the specific rate of enzyme production tended to increase.

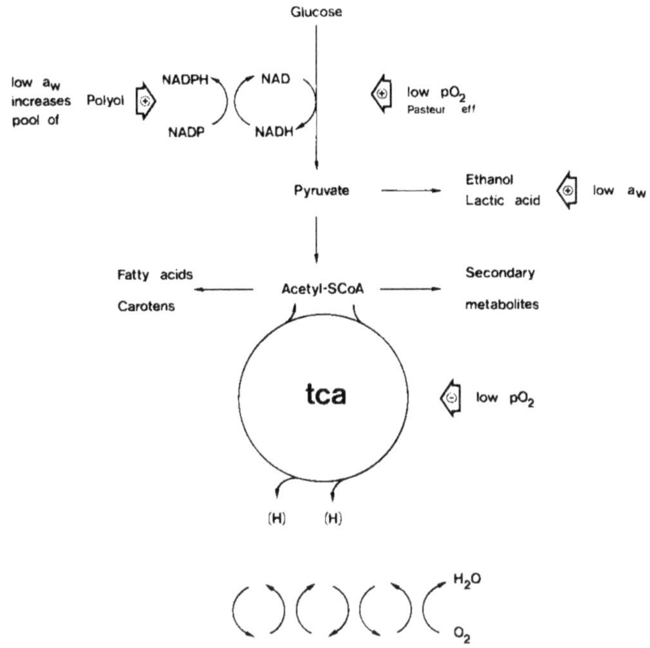

FIGURE 1. Schematic presentation of the metabolism of an aerobic organism. By arrows are indicated some of the effects that may appear upon changes in water activity or pO_2 in the environment of the cells.

Limited oxygen supply to the immobilized cells will change their metabolism. Some of the possible routes are indicated in Figure 1, which can be used when interpreting the product pattern of immobilized cells. The fact that immobilized plant cells tend to produce more secondary metabolites (see Chapter 3, Volume I) than free cells can also be interpreted along the lines presented in Figure 1.

Reduced cell growth in the immobilized state can be regarded as a positive effect, since cell growth in many cases can cause problems such as cell leakage and breakage of the support. The reason why cell growth is slower is not obvious, but based on the scheme in Figure 1, a general theory may be formulated.[16-18] It is a well-known fact that microorganisms stop growing at low water activity.[19]

Esener et al.[20] found a correlation between the water activity and the maximal growth rate for *Klebsiella aerogenes*. Concomitantly with a decrease in growth rate, a marked increase in the fraction of substrate used for maintenance of the metabolism was observed. Furthermore, when placing the cells in a milieu of lowered water activity, the cells compensated for the external environment by internally increasing their osmotic pressure. This could be done by production of specific metabolites like glycerol,[21,22] amino acids,[23] betaine,[24] etc.

It was furthermore found that increases in the intracellular contents of polyalcohols, like glycerol or mannitol, switched the metabolism in certain directions. Thus it was observed that *Dunaliella*, when growing at high salt, contained an intracellular concentration of glycerol of 7 M, and furthermore that the metabolism of these cells was switched, in the sense that NADH was converted to NADPH via a specific glycerolcycle.[21] Similar cycles have been found in fungi imperfectii.[25]

A change from NADH to NADPH leads to a change in the availability of redox-equivalents; from catabolism to anabolism. *Dunaliella* turned out to have improved yields of β-caroten at high salt conditions (see Figure 1).

Reduced water activity is a reality in immobilized cell preparations. It has been demonstrated, for example by Holzberger and Margalith,[26] that at 5g/ℓ sodium alginate a marked effect on the metabolism of yeast cells can already be observed. The cells showed a higher productivity (20 to 25% higher) compared to free cells. During the immobilization process, however, a high concentration of Ca^{2+} caused other disturbances, so that this effect was no longer clearly observed.

The influence of polymers on the water activity is not pronounced at low concentrations. However, it seems realistic to assume that a nonhomogeneous distribution of polymer (electron micrographs on sections of immobilized cell preparations) makes it possible for specific microenvironments to occur.

In a study of *Saccharomyces cerevisiae* adsorbed or covalently attached to porous glass, Navarro and Durand[27] found that the adsorbed cells showed a changed metabolic behavior in comparison to what was known to occur in free solution. In their fermentation of glucose to produce ethanol, a theoretical yield has been calculated based on empirical data of 0.51 g/g, assuming metabolism followed the established pathways. However, with the adsorbed cells yields of up to 53.6% were observed (an increase of about 7%. This increase was accompanied with a decrease in carbon dioxide productivity. Even if the mechanisms are not yet clearly understood, it seems reasonable to assume that the adsorbed cells underwent a change in their metabolic patterns.

Another observation of interest when discussing the physiology of immobilized cells is the changed pattern of division for covalently immobilized yeast cells.[28] It was seen that these cells, when multiplying, did not give significant leakage of progeny into the medium. From electron microscopic investigations it was determined that cell division proceeded without separation of the daughter cells; the formation of chain-like filaments indicated a certain polarization of budding.

In summary, the physiology of immobilized cells is a virgin area for investigators. Some indications are given above, but if a proper understanding of the metabolic behavior of immobilized cells is to be gained in order that a basis be set for designing immobilization systems for specific tasks, much more research is needed on this subject.

IV. CULTIVATION WITHIN THE SUPPORT

In applications where adsorbed cells are used in waste water treatment, the bed of support material is first inoculated with the organisms which are then allowed to multiply and fill the surface and/or available space. This kind of adhesive fermentation can be applied to many systems.

However, in more well-defined systems with adsorbed or entrapped monocultures or well-defined mixed cultures, cell growth can be of interest (Figure 2). In the preparations of carrageenan-immobilized yeast only a low content of cells was present in the immobilization step when the gel beads are formed. The cells were then allowed to grow.[29] An advantage of such a procedure is that the cells grow to a very high cell density near the surface of the gel, whereas the interior of the gel particles are almost empty. This results in beads having the cells in a layer where it can be expected that they come in contact with substrate, and that only very few cells in regions where no substrate will appear, overall, the cells will have a high effectiveness factor.

Another situation when postimmobilization cell growth is valuable occurs when vegetative cells of the organisms cannot stand the conditions during the immobilization. In such cases, if spores are available, they can be immobilized instead. The preparation then has to be activated to obtain vegetative cells. In the case of *Clostridium acetobutylicum*, which is a strictly anaerobic organism, it turned out to be a great advantage

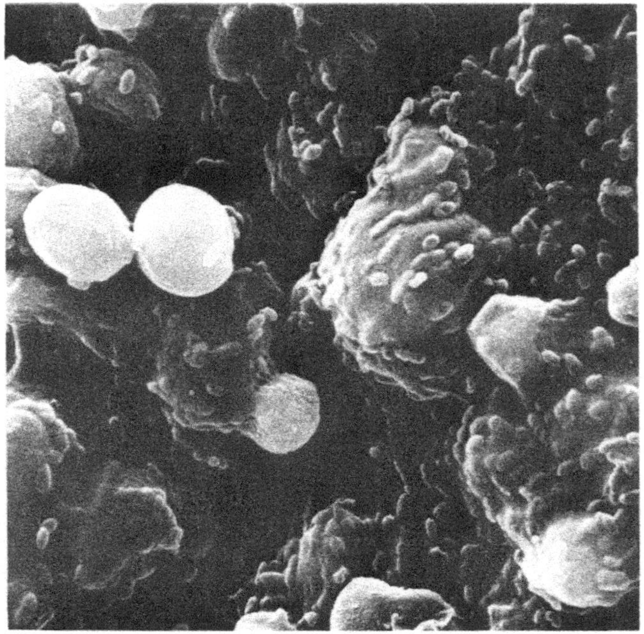

FIGURE 2a. Scanning electron micrograph of a coimmobilized algae-bacteria preparation after 21 hr in the reactor. The gel beads were fixed in 2% glutaraldehyde in 0.2 *M* sodium cacodylate buffer for 1.5 hr and postfixed in 2% osmium tetroxide in the same buffer for 1 hr. The beads were then dehydrated in an ethanol series and transferred to freon. Then the beads were dried using the CO_2 critical point drying technique and coated with gold-palladium. Magnification × 2400. (Photo taken by P. Adlercreutz.)

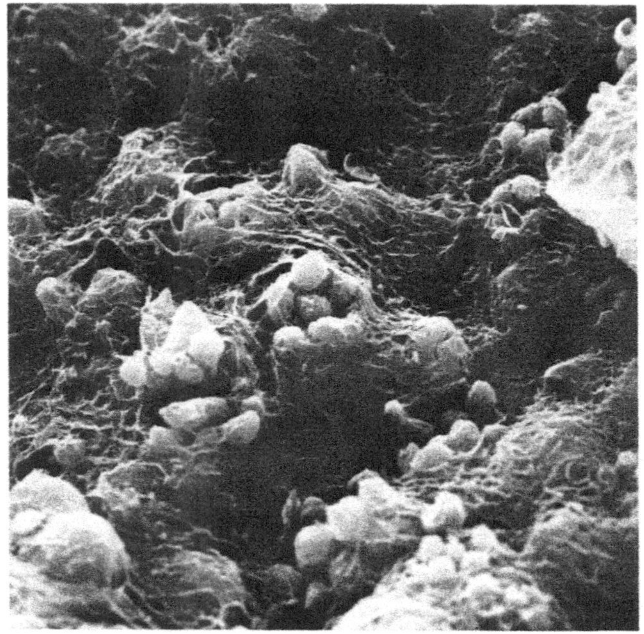

FIGURE 2b. Scanning electron micrograph of a coimmobilized algae-bacteria preparation after 70 hr in the reactor. The gel beads were briefly fixed in 2% glutaraldehyde in 0.2 *M* sodium cacodylate buffer for 15 min. Then the beads were cooled in melting freon 22 and fragmented. The fragments were freeze-dried and coated with gold-palladium. Magnification × 1200. (Photo taken by P. Adlercreutz.)

Table 2
IMMOBILIZED SPORES

Microorganism	Immobilization method	Comment	Ref.
Aspergillus niger	Adsorbed in controlled pore inorganics	Mycelium forming	32
Aspergillus oryzae	Adsorbed on ECTEOLA-cellulose		33
Aspergillus wentii			33
Clostridium acetolycetylium	Ca-alginate	Strictly anaerobic	30
Curvularia lunata	Ca-alginate	Low yield when negative cells were immobilized	31
Curvularia lunata	Photocross-linkable		34
Curvularia lunata	Entrapment in (1) photocross-linked resin (2) polyuretane (3) k-carrageenan (4) calcium alginate (5) agar		35
Penicillium chrysogenum	Adsorbed in controlled pore inorganics	Mycelium forming	32
Penicillium roquefortii	Adsorbed on ECTEOLA-cellulose		33
Streptomyces olichromogenes	Adsorbed in controlled pore inorganics	Mycelium forming	32

to use spores during the immobilization step[30] (see Chapter 3, Volume II). As well, in cases where there are labile structures in the vegetative state, spore-immobilization can be a good alternative.[31] Table 2 lists some examples where spores have been immobilized.

Using porous ceramics and porous glass, Messing et al.[32] investigated the relationships between the accumulation of stable and viable biomass and the pore morphology of dimensionally stable inorganic carriers. Adsorption as well as covalent coupling of cells was investigated. The results indicate that a pore size of 3 to 5 times that of the individual cells was optimal, as judged from the amount of biomass that was bound per g of support used. They studied microbes that reproduced by fission *(Escherichia coli, Serratia marcescens,* and *Bacillus subtilis),* those reproducing by budding *(Saccharomyces cerevisiae,-Saccharomyces amurcae)* and also microbes that exhibit mycelial growth and produce spores *(Aspergillus niger, Streptomyces olivochromogenes,* and *Penicillium chrysogenum).* In all cases it was found that a pore size of up to 5 times that of the individual cells was optimal. It should furthermore be stressed that the size of spores makes it physically feasible to introduce the mycelia-forming species into the pore-network of the supports used.

In many applications, postimmobilization cell growth is a great problem since cells leak from the sorbent and contaminate the product stream. However, the phenomenon can also be turned into a positive phenomenon in the sense that the bed of immobilized cells can be used to produce cellmass recovered in the effluent. Durand[36] demonstrated that immobilized synchronized yeast cells liberated cells into the effluent simultaneously, thereby causing a periodic pattern of time dependent cell number in the effluent.

The immobilization of spores may be regarded as a specific case of immobilized resting cells. The reasoning behind this statement is that both the cells and the spores are inactive with regard to division, whereas they still use certain operational enzyme functions. By changing the composition of the medium, it is possible to change both forms from a "resting" state to an "active" state. This is further discussed in the activation section.

V. PERMEABILIZATION

The kinetics of the processes catalyzed by immobilized cells is greatly influenced by diffusion. For free cells the transport of substances across the cell membrane is often a severe barrier, unless specific transport mechanisms are available. When immobilizing the biocatalyst, several additional barriers are added. The case of oxygen is illustrated schematically (Figure 1 in Chapter 3, Volume II). Diffusion through the Nernst diffusion layer can be influenced by either increasing the concentration gradient over the layer, i.e., increasing the bulk concentration of substrate, or by reducing the thickness of the layer. This can be achieved by increasing the stirring speed.

Internal diffusion within the gel particle is very difficult to change. However, by choosing proper polymer support it might be possible to achieve an enrichment of the substrate within this microenvironment. Such effects have been shown to be important for the kinetic behavior of immobilized enzymes.[37,38] The method must of course also be applicable to cells, but very little research has currently been done in this area.

The next great diffusion barrier is the cell membrane. Normally, a rather limited number of molecules have specific channels for entry, translocases or permeases available in order to gain access into the cell. Most molecules used in biotechnological conversions, however, are not among these selected substances. The cell-membrane barrier is then a severe hindrance to diffusion, and as such it impedes the catalytic processes to a high extent. Much effort has been spent to permeabilize the cell membranes in order to facilitate exchange between the interior and the exterior of the cell. However, if the cell membrane structure is loosened up, low molecular weight metabolites will tend to dissappear from the cell.

If the cell membrane is more severely destroyed, even proteins will leak. This method has in fact been used to study the topography of the cell since enzymes nearest to the membranes tend to escape first. Thus it is a delicate task to permeabilize a cell; it must be opened to such an extent so as to allow passage of substrate from exterior to interior but concomitantly still keep enzymes, and preferentially also coenzymes, within the cell. Several different approaches have been examined. The first cells to be studied were used for one-step conversions where no enzyme sequence and no coenzymes were involved. Several examples on such permeabilized systems are listed below. When operating with one-step conversions permeabilization may be useful both for facilitating substrate supply, and also for making it possible for reactants in competitive reactions to escape. The initial permeabilization efforts on immobilized cells were performed using metal ions. Today it is more conventional to use detergents or organic solvents. In some cases it seems that the immobilization process *per se*, or the microenvironment offered upon immobilization,[40,41] causes altered patterns in permeability.

Permeabilization can be performed by treatment with EDTA,[42] organic solvents,[43] detergents[44,45] and in specific cases, by other methods. Such one case is the use of heat treatment (60°C) to improve the permeability of the cell.[46] Freeze-thawing of cells has also been a method of choice. Kawabata and Demain[47] studied the synthesis of pantothenic acid by immobilized *Escherichia coli*. It soon turned out that permeabilization of the cells improved the overall catalytic activity. Freeze-thawing proved to be a very efficient method of getting conditions for keeping the enzymes within the cell and getting a metabolic response to the externally added ATP.

In the initial developmental stages of permeabilization the cells often died. In many cases that was of minor importance as long as the expected reaction operated according to plan, however, if coenzyme regeneration and other sequential reactions needed to be utilized, a rather high degree of the cells natural state needed to be preserved.

One point that still needs to be clarified is whether or not (or to what extent) permeabilized cells are viable. This must take place in parallel with an evaluation of different potential permeabilization methods.

VI. ACTIVATION

There are several reports describing procedures to improve the activity of immobilized cells. Permeabilization is an instant method which is used directly on the preparation to improve the initial activity. However, in many cases, cells show a sufficiently high initial activity and do not need activation until their initial activity falls off with time. In such cases the cells can be boostered with medium so that protein synthesis regenerates and restores the normal levels of essential enzymes and coenzymes of the immobilized preparation.

Activation of this type was first demonstrated by Mosbach and Larsson[7] studying *Curvularia lunata* entrapped in polyacrylamide. The cells could be activated under storage conditions by incubating the gel in a corn-steep-liquor medium. The technology was further improved on entrapped *Arthrobacter simplex*. Using this latter organism a remarkable improvement in catalytic power was observed. An increase in catalytic capacity of the cells by a factor of around 10 was observed and the time needed for a batch process to go to completion was reduced from initially 17 hr to 2 hr. It should in this context be noted that the enzyme as such is induced. Perhaps this kind of activation is especially suited for induced enzymes.

It was first noted during such activation experiments that as cell multiplication in the immobilized state occurred, a subsequent leakage of cells appeared in the medium.

Today activation is tried more or less on a routine basis. In cases when growth related processes are performed by the immobilized preparation, a dilute growth medium may be present during the whole operation. However, from the point of downstream processing it is advantageous to use clear product streams with intermittent periods of activation.

The mechanism behind activation has been under debate. Was the increased activity due to synthesis of new enzyme molecules within the old cells or was it a result of cell-proliferation resulting in new active cells?

Microscopic observations have often been used as well as dry or wet weight determinations in order to answer this question. In the case discussed above concerning *Arthrobacter simplex*, it was found from a microscopic examination of saphranine stained cells that the cell number had improved substantially. The same result was obtained when dry or wet weight determinations were measured. To ensure that the activation (at least to a major part) was ascribed to the generation of new cells, activation of entrapped cells was performed with growth-inducing medium and, in a parallel experiment, with medium supplied with 25 mg/mℓ of benzylpenicillin and 500 μg/mℓ of chloramphenicol, respectively. These two antibiotics, are known to inhibit cell wall synthesis and protein synthesis, respectively. Their use influenced the outcome of the activation process severely (Figure 3). The conclusion, at least in the case of *Arthrobacter simplex* must then be that the activation is to a major part directly coupled to cell proliferation within the gel.[49]

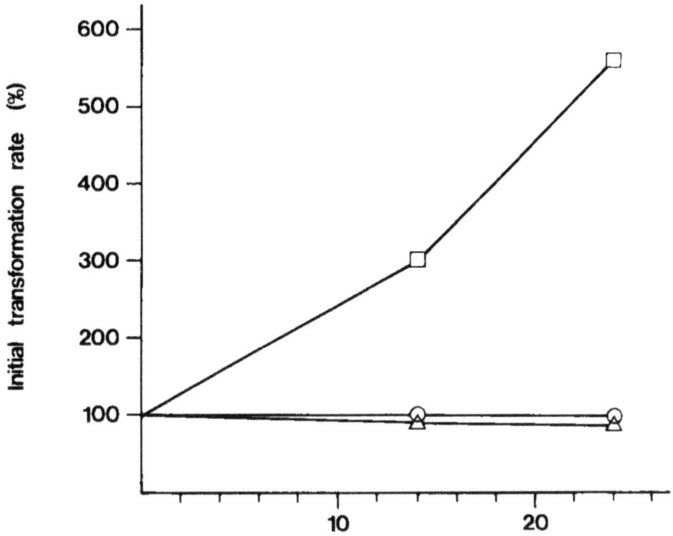

FIGURE 3. Influence of antibiotics on the activation by growth medium of *Arthrobacter simplex*. (□) No antibiotics added; (○) benzylpenicillin 25 mg/mℓ; (△) chloramphenicol 500 μg/mℓ. Activity of freshly prepared gel was set at 100%. (From Ohlson, S., Larsson, P.-O., and Mosbach, K., *Biotechnol. Bioeng.*, 20, 1267, 1978. With permission.)

There may be several reasons for a decrease of activity of immobilized cells. Denatured enzymes is one, another being lost coenzymes. This latter factor may be important, especially when permeabilized cells are used. It has thus been shown that additions of artificial electron acceptors may improve the expressed catalytic ability of native cells,[50] as well as of permeabilized ones.[51]

VII. SIDE REACTIONS

One crucial point, often raised especially during the initial period when immobilized cells became an alternative to the use of immobilized enzymes, was that of side reactions. How do we know that the cell, with a very broad spectrum of metabolic potentials, will make just the expected substance and nothing else, or even continue to metabolize the product even further?

Initially, when dead cells were used for carrying out one or a few specific steps this problem could, in many cases, be controlled. Traditional fermentation technology supplied the already developed procedures. A basic philosphy had been to use mutants that were unable to metabolize the product further. In many cases specific experimental procedures had to be developed to selectively eliminate postproduct enzyme activities while keeping the preceding steps in a metabolic sequence intact.

An example of this was presented by Yamamoto et al.[52] who developed a process for the production of urocanic acid. *Achronobacter liquidum* contains the enzyme L-histidine ammonialyase which converts L-histidine to urocanic acid. Unfortunately the cells also contained urocanase which transforms urocanic acid into imidazolone propionic acid. To eliminate this latter activity the cells were heated to 70°C for 30 min before they were immobilized.

Another example of a discontinued enzyme process was demonstrated on *Brevibacterium ammoniagenes,* which converts fumarate to the desired product L-malic acid and then further on to succinic acid. Permeabilization of the cells, either by treatment with acetone or with detergents, caused a suppression of succinic acid formation. This was interpreted to be due either to a difference in enzyme stabilities to these treatments, or to a severe loss of essential coenzymes for the latter process.[45]

Both those examples have utilized dead cells. They were included here since it is much easier to demonstrate which activities have to be eliminated. For cells with their metabolic systems intact, the situation is rather complicated. First, for resting cells, many potential metabolic routes exist. To be able to utilize these cells, certain precautions must be set. One-step conversions by viable cells can, in many cases, be performed by using a mutant strain that will stop at the desired product. In some cases products are end-products and then this problem of specificity does not occur.

The greatest problem occurs when processes are growth related, since then the substrate that shall be converted to the desired product can also be involved in the cell metabolism. This situation will be reflected by a net increase in cell mass at the expense of product. In these cases it may be possible to manipulate the metabolism towards the desired direction. The paragraph on the physiology of the immobilized cells gives some guidelines to follow in this matter.

Another method may be the use of selective metabolic inhibitors.[53] By adding azide to a fermentation medium it was possible to produce ethanol under aerobic conditions. In the normal situation all energy would have gone into cell mass-production, but with azide present good yields of ethanol were obtained. A similar approach was taken by Disping and Rehm,[54] when using immobilized *Saccharomyces cerevisiae* to produce glycerol from glucose, by the addition of sodium sulfite.

Very few papers report on byproducts. Often the author only provides figures for the yields of the desired product. However, Vieth and Venkatsubramanian[55] in their report on citric acid production by *Aspergillus niger* attached to collagen, found with chromatographic analysis of the medium that side reactions had occurred, leading to the production of isocitric acid, oxalic acid, and trace quantities of gluconic acid. Isocitrate was the major contaminant, as high as 15 to 20% of the final product yield.

VIII. APPLICATIONS

At present it is difficult to know which reactions have been brought to industrial scale, which are at a pilot plant scale and which are still at a laboratory scale. Most reports concern lab scale experiments and in a few cases concern themselves with the larger scales. A clear picture is difficult to get at this point in time.

Among the reactions mentioned in this chapter and those discussed in relation to mild immobilization methods (Chapter 2, Volume I) are many with great potential for industrial applications.

The easiest systems to bring to an industrial scale are those where only small volumes of high price-product are produced. When low-priced bulk-chemicals are to be produced, much more rigorous restrictions concerning economy and technical aspects of scale up have to be laid down. These latter facts will perhaps delay the exploitations to some degree.

An area where immobilized viable cells should have a great potential is when they offer certain advantages or even make it possible to perform processes that otherwise are not possible.

In some cases when microbial cells are to be used for a conversion process experimental conditions far from those optimal for the cell must be chosen. This is the case for the food area, where pH, ionic strength, temperature, and presence of specific metabolites cannot be changed.

In the treatment of nitrate-containing water for human consumption a situation like the one described above is at hand. Here no additives should be added to the water, but it has been shown that the cells in order to be able to convert nitrate to nitrogen, also need a carbon source.[56]

Otherwise, no further additions are necessary. When using immobilized *Pseudomonas denitrificans,* aspartate was found to be the best carbon source but ethanol was acceptable as well.[57] The amount of carbon source should be sufficient to make the cells reduce the nitrate content but no more.

Here, however, another problem appeared. Leakage of cells from the alginate immobilized preparation occurred. To reduce this leakage stabilization of the gel[58] as well as utilization of membrane technology[59] was tried. In the latter case an efficient recovery of cells was achieved during an an initial phase before a lower level of cell-leakage appeared.

Another system with defined reaction parameters is in the production of wine. Here pH, salt concentration, and temperature cannot be freely altered. Furthermore, wine contains high concentrations of ethanol and the preservative SO_2, making cell growth very low. Under such circumstances it may turn ideal to use immobilized viable cells since in such a preparation a dense cell population exists that will grow slowly, which in turn reduces the content of cells in the effluent. The high cell density within the beads will enable the malolactic fermentation to occur at higher rates than this possible with free cells. *Leuconostoc oenos,* immobilized in alginate was used in this study.[60]

IX. FERMENTATIVE PRODUCTION OF ETHANOL USING IMMOBILIZED MICROORGANISMS

Among the fermentative processes studied based on immobilized viable cells, production of ethanol is the most popular.[61] This may be so because of the need for other liquid fuels than petrochemical products, but another very important factor has been that ethanol fermentation by *Saccharomyces cerevisiae* is a very cheap and easily operated model reaction. Yeast cells must be viable to be able to perform the fermentative conversion of glucose to ethanol and carbon dioxide. To keep the cells in condition for continuous fermentation, media for cell growth must be present. Chibata et al.[62] demonstrated that when the immobilized cell preparation was fed with only glucose productivity remained constant for 2 days and then gradually ceased. However if growth medium or a reduced growth medium was added to the glucose feed a continuous process could be run for 3 months. In another experiment the possibility of feeding the cells with growth medium intermittently[63] had been demonstrated.

As is seen from Chapter 2, Volume I, there are numerous reports on the immobilization of *S. cerevisiae,* and it seems that operational immobilization methods are at hand. The main bottleneck in the process is product inhibition. The endproduct, ethanol, inhibits the activity of the cells.

When operating with entrapped cell preparations mass transfer restrictions may also appear. The production of carbon dioxide may cause severe disturbances. To reduce this effect it may, under certain conditions, be worthwhile to operate the system under pressure to improve the solubility of CO_2 in the medium. If this is not possible it is favorable to operate with cells arranged in thin layers, either adsorbed[64-66] or in beads with an asymmetric distribution of entrapped cells.[29]

One of the main issues being dealt with when producing ethanol by immobilized cells is the productivity, but very little is said about final concentration.

This is an important factor to consider when evaluating the economy of the fermentative production of ethanol. Since distillation is the main method used for separation, it is a neccessity to keep the ethanol concentration in the feed line to a distillation process at as high level as possible. Low levels of ethanol, cause large costs in the energy needed for distillation and will thus make the process uneconomical.

Table 3

Organism	pH	Temp (°C)	Substrate	Ethanol conc. (w/v %)	Productivity g ethanol/h · l	Ethanol yield g ethanol/ g substrate	Comments	Ref.
Saccharomyces cerevisiae	5.2-4.8	30	Molass with 20% sugar	7.8%	1.1 g/h · l	0.39	Batch	67
		35	Glucose sucrose 10%	4.5%	7 g/h · l	0.45	Continuous	68
	5.4		Molass 10-11%	5.7-5.9%	∼50 g/ℓ · h	0.51	Continuous with recirculation of cell mass (flockulation)	69
	3.5-4.0	35	Glucose sucrose 50%		82 g/ℓ · h	0.47-0.49	Continuous, recirculation, vacuum	70
			Sucrose		22 g/ℓ · h	0.53	Continuous, recirculation, (adsorption)	64
		30	Glucose 12%	4.8%	53 g/ℓ · h	0.37	Adsorption dilution rate 1.1 h^{-1}	65
		30	Glucose 12%	4.3%	62 g/ℓ · h	0.51	1.44 h^{-1}	65
Zymomonas mobilis			16-18% glucose	>6.0%	152 g/ℓ · h		Adsorbed	66
			16.5% glucose		40.5 g/ℓ · h		Cell recycle	70
	5.0	30	Glucose 10%	4.9%	7 g/ℓ · h	0.49	Continuous	71
	5.0	30	Glucose 10%	5%	3.3 g/ℓ · h	0.50	Batchwise	72
			Glucose 20%	10%	3.3 g/ℓ · h			72

The main problem is thus to eliminate the ethanol formed from the site of reaction, but to keep its concentration in the effluent high. Table 3 gives some examples of different possible solutions to this dilemma. As is seen from the data, it is possible to achieve high productivity, but normally at the expense of the final concentration of ethanol in the medium.

X. CONCLUDING REMARKS

There are promising potentials in using immobilized viable cell technology since this approach makes it possible to operate with high cell densities and high dilution rates at high operational stability.

Practical large-scale applications are still rare — at least those that are officially reported. The use of mixed cultures in waste water treatment (Chapter 5, Volume II) and the process for conversion of ethanol to acetic acid by immobilized *Acetobacter* (especially used in former days) are among the few in operation.

As judged from the large number of reports on immobilized viable cells which have been used for a broad spectrum of bioconversions, an interesting future is foreseeable. The technology for immobilizing cells is well established today, even for labile structures. What now is needed is an input from the engineers in order to evolve better reactor designs, and to improve the quality of process engineering and down-stream processing.

REFERENCES

1. **Freeman, A., Blank, T., and Aharonowitz, Y.,** Protein determination of cells immobilized in cross-linked synthetic gels, *Eur. J. Appl. Microbiol. Biotechnol.*, 14, 13, 1982.
2. **Mattiasson, B., Larsson, P.-O., and Mosbach, K.,** The microbe thermistor, *Nature (London)*, 268, 519, 1977.
3. **Kierstan, M. and Bucke, C.,** The immobilization of microbial cells, subcellular organelles and enzymes in calcium alginate gels, *Biotechnol. Bioeng.*, 19, 387, 1977.
4. **Marcipar, A., Cochet, N., Brackenridge, L., and Lebeault, J. M.,** Immobilization of yeasts on ceramic supports, *Biotechnol. Lett.*, 1, 65, 1979.
5. **Kennedy, J. F., Barker, S. A., and Humphreys, J. D.,** Microbial cells living immobilized on metal hydroxides, *Nature (London)*, 261, 242, 1976.
6. **Klein, J., Hackel, U., and Wagner, F.,** Phenol degradation by *Candida tropicalis* whole cells entrapped in polymeric ionic networks, in *Immobilized Microbial Cells*, ACS Symp. Series, Vol 106, American Chemical Society, Washington, D.C., 1979, 101.
7. **Mosbach, K. and Larsson, P.-O.,** Preparation and application of polymer entrapped enzymes and microorganisms in microbial transformation processes with special reference to steroid 11-β-hydroxylation and Δ'-dehydrogenation, *Biotechnol. Bioeng.*, 12, 19, 1970.
8. **Sommerville, H. J., Mason, J. R., and Ruffel, R. N.,** Benzene degradation by bacterial cells immobilized in polyacrylamide gel, *Eur. J. Appl. Microbiol.*, 4, 75, 1977.
9. **Ohlson, S., Larsson, P.-O., and Mosbach, K.,** Steroid transformation by activated living immobilized *Arthrobacter simplex* cells, *Biotechnol. Bioeng.*, 20, 1267, 1978.
10. **Kokubu, T., Karube, I., and Suzuki, S.,** α-amylase production by immobilized whole cells of *Bacillus subtilis*, *Eur. J. Appl. Microbiol. Biotechnol.*, 5, 233, 1978.
11. **Suzuki, S. and Karube, I.,** Production of antibiotics and enzymes by immobilized whole cells, in *Immobilized Microbial Cells*, ACS Symp. Series, Vol 106, American Chemical Society, Washington, D.C., 1979.
12. **Chibata, I., Tosa, T., and Sato, T.,** Immobilized aspartase-containing microbial cells. Preparation and enzymic properties, *Appl. Microbiol.*, 27, 878, 1974.
13. **Goldstein, L.,** Kinetic behaviour of immobilized enzyme systems, *Methods Enzymol.*, 44, 397, 1976.
14. **Mattiasson, B.,** Biochemical applications and properties of immobilized multistep enzyme systems, in *Biomedical Applications of Immobilized Enzymes and Proteins*, Vol. 2, Chang, T. M. S., Ed., Plenum Press, New York. 1977, 253.

15. **Shinmyo, A., Kimura, H., and Okada, H.**, Physiology of α-amylase production by immobilized *Bacillus amyloliquefaciens*, *Eur. J. Appl. Microbiol. Biotechnol.*, 14, 7, 1982.
16. **Mattiasson, B., Adlercreutz, P., and Hahn-Hägerdal, B.**, Interphase Phenomena in Immobilized Cell Systems, Abstr. no 12. 19. Tutzing-Symposium der Dechema "Biologische Eigenshaften an Grenzflächen und ihre Bedeutung für die Biotechnologie" March 8-11, 1982.
17. **Mattiasson, B. and Hahn-Hägerdal, B.**, Microenvironmental effect on metabolic behaviour of immobilized cells, *Eur. J. Appl. Microbiol. Biotechnol.*, 16, 52, 1982.
18. **Hahn-Hägerdal, B. and Mattiasson, B.**, Shift in metabolism towards ethanol production in *Saccharomyces cerevisiae* using alterations of the physico-chemical environment, in Proc. 4th Symp. on Biotechnology in Energy Production and Conservation, Gatlinburg, Tenn., May 11-14, 1982.
19. **Scott, W. J.**, Water relations of food-spoilage microorganisms, *Adv. Food Res.*, 7, 83, 1957.
20. **Esener, A. A., Bol., G., Kossen, N. W. F., and Roels, J. A.**, Effect of water activity on microbial growth, in *Advances in Biotechnology*, Vol. 1, Moo-Young, M., Robinson, C. W., and Vezina, C., Eds., Pergamon Press, Toronto, 1981, 339.
21. **Ben-Amotz, A. and Avron, M.**, Glycerol and β-caroten metabolism in the halotolerant alga *Dunaliella*. A model system for biosolar energy conversion, *Trends Biol. Sci.*, 279, 1981.
22. **Edgely, M. and Brown, A. D.**, Response of xerotolerant and non-tolerant yeasts to water stress, *J. Gen. Microbiol.*, 104, 343, 1978.
23. **Measures, J. C.**, Role of amino acids in osmoregulation of nonhalophilic bacteria, *Nature (London)*, 257, 398, 1975.
24. **Galinski, E. A. and Trüper, H. G.**, Betaine, a compatible solute in the extremely halophilic phototrophic bacterium *Ectothiorhodospira halochloris*, *FEMS Microbiol. Lett.*, 13, 357, 1982.
25. **Hult, K. and Gatenbeck, S.**, Production of NADPH in the mannitol cycle and its relation to polyketide formation in *Alternaria alternata*, *Eur. J. Biochem.*, 88, 607, 1978.
26. **Holcberg, I. B. and Margalith, P.**, Alcoholic fermentation by immobilized yeast at high sugar concentrations, *Eur. J. Appl. Microbiol. Biotechnol.*, 13, 133, 1981.
27. **Navarro, J. M. and Durand, G.**, Modification of yeast metabolism by immobilization onto porous glass, *Eur. J. Appl. Microbiol.*, 4, 243, 1977.
28. **Jirku, V., Turkova, J., and Krumphanzl, V.**, Immobilization of yeast cells with retention of cell division and extracellular production of macromolecules, *Biotechnol. Lett.*, 2, 509, 1980.
29. **Chibata, I.**, Production of useful chemicals using cells immobilized with polyacrylamide and carrageenan, in *Enzyme Engineering*, Vol. 5. Weetall, H. H. and Royer, G. P., Eds., Plenum Press, New York, 1980, 393.
30. **Häggström, L.**, Immobilized cells of *Clostridium acetobutylicum* for butanol production, in *Advances in Biotechnology*, Vol. 2, Moo-Young, M. and Robinson, C. W., Eds., Pergamon Press, Toronto, 1981, 79.
31. **Ohlson, S., Flygare, S., Larsson, P.-O., and Mosbach, K.**, Steroid hydroxylation using immobilized spores of *Curvularia lunata* germinated *in situ*, *Eur. J. Appl. Microbiol. Biotechnol.*, 10, 1, 1980.
32. **Messing, R. A., Oppermann, R. A., and Kolot, F. B.**, Pore dimensions for accumulating biomass, in *Immobilized Microbial Cells*, ACS Symp. Series, Vol 106, American Chemical Society, Washington, D.C., 1979, 13.
33. **Johnson, D. E. and Ciegler, A.**, Substrate conversion by fungal spores entrapped in solid matrices, *Arch. Biochem. Biophys.*, 130, 384, 1969.
34. **Tanaka, A., Sonomoto, K., Hoq, M., Usui, N., Nomura, K., and Fukui, S.**, Hydroxylation of steroids by immobilized microbial cells, in 6th Enzyme Engineering Conf., Kashikojima, Japan, Sept. 20-25, 1981. Abstr. no 1-6.
35. **Sonomoto, K., Hog, M., Tanaka, A., and Fukui, S.**, Growth of *Curvularia lunata* spores into mycelial form within various gels, and steroid 11-β-hydroxylation by the entrapped mycelia, *J. Ferment. Technol.*, 59, 465, 1981.
36. **Navarro, J. M. and Durand, G.**, Modifications of the growth of *Saccharomyces uvarum* by immobilization on a solid carrier, *C. R. Acad. Sci. Ser. D*, 290, 453, 1980.
37. **Goldstein, L., Levin, Y., and Katchalski, E.**, A water-insoluble polyanionic derivative of trypsin. II. Effect of the polyelectrolyte carrier on the kinetic behaviour of the bound trypsin, *Biochemistry*, 3, 1913, 1964.
38. **Filippusson, H. and Hornby, W. E.**, The preparation and properties of yeast β-fructofuranosidase chemically attached to polystyrene, *Biochem. J.*, 120, 215, 1970.
39. **Katchalski, E., Silman, I., and Goldman, R.**, Effect of the microenvironment on the mode of action of immobilized enzymes, *Adv. Enzymol.*, 34, 445, 1971.
40. **Martin, C. K. A. and Perlman, D.**, Conversion of L-Sorbose to L-Sorbosone by Immobilized Cells of *Gluconobacter melanogenus* IFO 3293, *Biotechnol. Bioeng.*, 18, 217, 1976.
41. **Yamamoto, K., Sato, T., Yamashita, K., and Chibata, I.**, Continuous production of L-malic acid by immobilized *Brevibacterium ammoniagenes* cells, *Eur. J. Appl. Microbiol.*, 3, 169, 1976.

42. Leive, L., A nonspecific increase in permeability in *Escherichia coli* produced by EDTA, *Proc. Natl. Acad. Sci. USA*, 53, 745, 1965.
43. Weitzman, P. D. J., Behaviour of enzymes at high concentration. Use of permeabilized cells in the study of enzyme activity and its regulation, *FEBS Lett.*, 32, 247, 1973.
44. Shimizu, S., Morioka, H., Tani, Y., and Ogata, K., Metabolism of pantothenic acid in microorganisms. XI. Synthesis of coenzyme A by immobilized microbial cells, *Hakko Kogaku Zasshi*, 53, 77, 1975.
45. Chibata, I., Tosa, T., and Yamamoto, K., Industrial production of L-malic acid by immobilized microbial cells, in *Enzyme Engineering*, Vol. 3, Pye, E. K. and Weetall, H. H., Eds., Plenum Press, New York, 1978, 463.
46. Jirku, V., Turkova, J., Kuchynkova, A., and Krumphanzl, V., Modified hydroxyalkyl methacrylate gel as a support for immobilization of yeast cells, *Eur. J. Appl. Microbiol. Biotechnol.*, 6, 217, 1979.
47. Kawabata, Y. and Demain, A. L., Enzymic synthesis of pantothenic acid by *Escherichia coli* cells, in *Immobilized Microbial Cells* ACS Symp. Series, Vol. 106, American Chemical Society, Washington, D.C., 1979, 133.
48. Larsson, P.-O., Ohlson, S., and Mosbach, K., New approach to steroid conversion using activated immobilized microorganisms, *Nature (London)*, 263, 796, 1976.
49. Ohlson, S., Larsson, P.-O., and Mosbach, K., Steroid transformation by living cells immobilized in calcium alginate, *Eur. J. Appl. Microbiol. Biotechnol.*, 7, 103, 1979.
50. Stefanovic, V., Hayano, M., and Dorfman, R. I., Some Observations on the Δ'-Dehydrogenation of Steroids by *Bacillus sphaericus*, *Biochim. Biophys. Acta*, 71, 429, 1963.
51. Lee, B. K., Brown, W. E., Ryu, D. Y. and Thoma, R. W., Sequential 11-α-hydroxylation and 1-dehydrogenation of 16-α-hydroxycortexolone, *Biotechnol. Bioeng.*, 13, 503, 1971.
52. Yamamoto, K., Sato, T., Tosa, T., and Chibata, I., Continuous production of urocanic acid by immobilized *Achromobacter liquidum* cells, *Biotechnol. Bioeng.*, 16, 1601, 1974.
53. Hahn-Hägerdal, B. and Mattiasson, B., Azide sterilization of fermentation media. Ethanol production from glucose using immobilized *Saccharomyces cerevisiae*, *Eur. J. Appl. Microbiol. Biotechnol.*, 14, 140, 1982.
54. Bisping, B. and Rehm, H. J., Glycerol production by immobilized cells of *Saccharomyces cerevisiae*, *Eur. J. Appl. Microbiol. Biotechnol.*, 14, 136, 1982.
55. Vieth, W. R. and Venkatsubramanian, K., Immobilized microbial cells in complex biocatalysis, in *Immobilized Microbial Cells*, ACS Symp. Series, Vol 106, American Chemical Society, Washington, D.C., 1979.
56. McCarty, P. L., Beck, L., and St. Amant, P., Biological denitrification of waste waters by addition of organic materials, in 24th Annu. Purdue Industrial Waste Conf., Purdue Univ. Lafayette, Ind., 1969, 1271.
57. Nilsson, I., Ohlson, S., Häggström, L., Molin, N., and Mosbach, K., Denitrification of water using immobilized *Pseudomonas denitrificans* cells, *Eur. J. Appl. Microbiol. Biotechnol.*, 10, 261, 1980.
58. Nilsson, I., and Ohlson, S., Columnar denitrification of water by immobilized *Pseudomonas denitrificans* cells, *Eur. J. Appl. Microbiol. Biotechnol.*, 14, 86, 1982.
59. Mattiasson, B., Ramstorp, M., Nilsson, I., and Hahn-Hägerdal, B., Comparison of the performance of a hollow-fiber microbe reactor with a reactor containing alginate entrapped cells, *Biotechnol. Lett.*, 3, 561, 1981.
60. Gestrelius, S., A potential application of immobilized viable cells in food industry: malolactic fermentation of wine, in 6th Enzyme Engineering Conf., Kashikojima, Japan, Sept. 20-25, 1981.
61. Kolot, F. B., New trends in yeast technology — immobilized cells, *Process Biochemistry*, Oct/Nov, 2, 1980.
62. Wada, M., Kato, J., and Chibata, I., Continuous production of ethanol using immobilized growing yeast cells, *Eur. J. Microbiol. Biotechnol.*, 10, 275, 1980.
63. Lee, S. S. and Wang, H., Repeated fed-batch ethanol fermentation using immobilized yeast cells with on-line extraction system, in 4th Symp. Biotechnology in Energy Production and Conservation, Gatlinburg, TN, May, 11-14, 1982.
64. Moo-Young, M., Lamprey, I., and Robinson, C. W., Ethanol production by immobilized yeast cells in a packed tower, in VIth Int. Fermentation Symp., London, Ontario, Canada, July 20-25, 1980.
65. Daugalis, A. J., Brown, N. M., Cluett, W. R., and Dunlop, D. B., Production of ethanol by adsorbed yeast cells, *Biotechnol. Lett.*, 3, 651, 1981.
66. Arcuri, E. J., Continuous ethanol production and cell growth in an immobilized cell bioreactor employing *Zymomonas mobilis*, *Biotechnol. Bioeng.*, 24, 595, 1982.
67. Rose, D., Yeasts for molasses alcohol, *Process Biochem.*, March, 10, 1976.
68. Cysewski, G. R. and Wilke, C. R., Rapid ethanol fermentation using vacuum and cell recycle, *Biotechnol. Bioeng.*, 19, 1125, 1977.
69. Meyrath, J., Industrial production of ethanol by continuous fermentation, in Int. Microbiol. and Food Ind. Congress, Paris, 1979.

70. de Boks, P. A. and van Eubergen, G. C., Continuous ethanol production using cell recycle with a settler, *Biotechnol. Lett.*, 3, 577, 1981.
71. Lee, K. J., Tribe, D. E., and Rogers, P. L., Ethanol production by *Zymomonas mobilis* in continuous culture at high glucose concentrations, *Biotechnol. Lett.*, 1, 421, 1979.
72. Rogers, P. L., Lee, K. J., and Tribe, D. E., Kinetics of alcohol production by *Zymomonas mobilis* at high sugar concentrations, *Biotechnol. Lett.*, 1, 165, 1979.

Chapter 3

OXYGENATION OF PROCESSES INVOLVING IMMOBILIZED CELLS

S. O. Enfors and Bo Mattiasson

TABLE OF CONTENTS

I. INTRODUCTION

The application of immobilized cells has hitherto been mainly concerned with anaerobic processes, either by the use of strictly anaerobic organisms (see also Chapter 4, this volume) or by single nonoxygen requiring enzymatic reactions of the cell. One reason for this is, no doubt, the difficulty of reaching high enough oxygen transfer rates to the immobilized cells. Oxygen transfer causes considerable technical problems in ordinary fermentation processes with suspended cells and this problem is further amplified when the cells are immobilized. Oxygen is used in the cells for two main purposes. First as an electron acceptor in the respiration chain, in which case it is bound as water and is not incorporated in the biomass, and secondly as a substrate for the oxygenases, in which case one or both of the oxygen atoms will be bound to the other substrate molecule and thus appear in the biomass.

Quantitatively, the most important utilization of oxygen is the respiration process. Here approximately equal amounts of carbon dioxide will be formed to the amounts of oxygen consumed. The carbon dioxide will be subject to similar transport mechanisms, but in the reverse direction to that of oxygen. The desorption of carbon dioxide from immobilized cells may be as important a problem as supplying the cells with oxygen, since many microorganisms are markedly inhibited by elevated carbon dioxide concentrations.[1]

The problem of carbon dioxide elimination is not unique for aerobic processes and it will not be further discussed here. A theoretical treatment of this problem has recently been presented by Krouwel and Kossen.[2]

II. MECHANISMS OF OXYGEN TRANSFER

The transport of oxygen from air to the immobilized cell is composed of a series of reactions which impose different resistances to the overall process. This is visualized in Figure 1.

A. The Gas-Liquid Transfer of Oxygen

In common fermentation processes step 5 of Figure 1 is eliminated and step 4 may represent the diffusion through the stagnant liquid film surrounding the cell. Each step involves a resistance to the transport of oxygen which can be written 1/k, where k is the mass transfer coefficient.

Several scientists have investigated the relative importance of the different steps of the gas to cell transfer of oxygen and found that the diffusion through the stagnant liquid film around the air bubble (step 2) is the rate limiting reaction.[3] According to Fick's first law of diffusion this process can be written as:

$$N = D \cdot a \cdot dC/dX \qquad (1)$$

where N = the flux of oxygen, D = the diffusivity of oxygen through the stagnant liquid film, a = interfacial area, and dC/dX = the oxygen concentration (C) gradient through the film (X). Integrating Equation 1 within the limits of the film gives

$$N = \frac{D \cdot a}{x} (C^* - C) \qquad (2)$$

where C* and C are the oxygen concentrations at the interface and in the bulk liquid, respectively. Since the individual parameters D, a, and X are difficult to determine

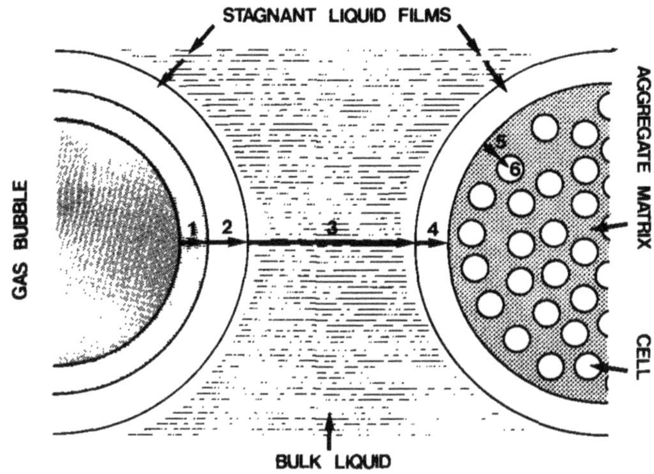

FIGURE 1. A schematic representation of the series of transport reactions from the gas phase to the immobilized cell. All steps, except step 3, are mainly diffusion controlled. The oxygen transport in the bulk liquid is achieved by hydromechanical means.

experimentally they are mostly combined to a volumetric overall oxygen transfer coefficient k_La (h^{-1}) and the oxygen transfer rate in the reactor (OTR, mmol/ℓ-h) is written as:

$$OTR = k_L a (C^* - C) \tag{3}$$

where k_L is the overall oxygen transfer coefficient (cm/h), a is the gas-liquid interphase area per unit volume (cm^2/cm^3), C^* is the oxygen concentration at the gas-liquid interphase (mmol/ℓ) and C is the actual oxygen concentration of the medium.

k_La is frequently used to characterize the oxygen transfer capacity of a reactor. Alternatively, the maximum oxygen transfer rate, i.e., $k_L \cdot a \cdot C^*$ (mmol/ℓ·h), can be used. The latter parameter is obtained by means of the sulfite oxidation method,[4] while the gassing out method[5] or the dynamic method[6] give k_L a values. When C^* is known, these parameters can be calculated by using Equation 3. Typical data for the OTR of some equipments is given in Table 1. There are three main ways to control the OTR in a reactor. The value "a" can be increased by increasing the agitation or the air flow rate. This also means that C^* will increase since the mean oxygen content of the ascending air bubbles will increase. A third way is to increase C^* directly by increasing the partial pressure of oxygen in the inlet air either by increasing the total air pressure or by mixing pure oxygen into the air. k_L may to some extent depend on the agitation but it is mainly determined by the medium composition and the temperature and it is therefore not subject to control. An extensive review on the subject of gas-liquid oxygen transfer was published by Tsao and Lee.[7]

B. The Liquid-Cell Transfer of Oxygen

In common fermentations with suspended cells the most important process controlling oxygen transfer to the cells is the diffusion of oxygen through the gas-liquid interphase. However, when cells are immobilized in a matrix, a second series of diffusion steps are added (steps 4 to 6 in Figure 1), diffusion through the stagnant liquid film surrounding the cell aggregate, diffusion through the matrix of the aggregate, and finally, diffusion through the cell wall. These three steps are illustrated in Figure 1.

Table 1
TYPICAL VALUES OF k_La AND MAXIMUM OTR FOR SOME TYPES OF CULTIVATION APPARATUS

	k_La (h^{-1})	Max OTR[a] (mmol/ℓ·h)
Erlenmeyer flasks on shaker	20-200 [b]	5-50
Air lift fermenter	400-800[c]	100-200
Conventional stirred fermenter	800-2400[d]	200-600

[a] Derived from K_La under the assumption that C^* is 0.25 m*M*, i.e., the solubility of oxygen in water at 26.8°C and equilibrium with 1 atm of air.

[b] Most important parameters: volumetric ratio medium: flask, shaker rate, and amplitude (orbit diameter).

[c] Most important parameters: air flow rate, ratio reactor height: diameter.

[d] Most important parameters: stirrer speed, air flow rate.

The flux of oxygen (N) through each step can be described by Fick's first law of diffusion:

$$-N = D(dC/dX) \tag{4}$$

where D = diffusivity of oxygen and C = oxygen concentration at distance X from the surface of the diffusion barrier.

Yoshida et al.[8] summarized the work of other researchers on the liquid film resistance of single cells. This was generally considered to be negligible. Experimental work by Yoshida's group on oxygen diffusion to large mycelial pellets also showed that the liquid film around the cell aggregate constituted a minor resistance to the oxygen transfer under ordinary culture conditions. However, Sada et al.[9] found when working with polyacrylamide beads containing entrapped urease and magnetite, that movement of the beads in a magnetic field efficiently reduced the liquid film resistance and thus improved the transfer of substrate to the outer surface of the matrix.

The transport of oxygen within aggregates of cells can be partially controlled with bulk flow if the aggregate is of low density (like diffuse mycelial pellets). In more dense pellets, however, the oxygen flux is mainly controlled by molecular diffusion and this should be the case also in aggregates of gel-entrapped cells. A discussion on oxygen diffusion coefficients in microbial aggregates is presented by Ngian and Lin who compared different results reported in literature.[10] As a rule oxygen diffusivity is somewhat lower in microbial aggregates than in water.

Little information is available on the importance of the diffusional resistance to the oxygen flow exerted by the cell, however, Langmuir[11] observed that the K_S for oxygen of respirating microorganisms was proportional to the cell size and that cell-free extracts of the largest cells exhibited similar K_M as that of the smallest cells in spite of the 100-fold higher K_S for the intact large cells. This might indicate that the last step of oxygen transfer is also markedly restricted by diffusional resistance (see Table 2).

Table 2
k_r-VALUES OF SOME MICROORGANISMS AND THEIR CELL-FREE EXTRACTS[a]

Organisms	Ø (μM)	k_r (% Air Sat)
Micrococcus candicans	0.5	0.004
Aerobacter aerogenes	0.6	0.01
Escherichia coli	0.6	0.01
Serratia marcescens	0.7	0.01
Azotobacter indicum	1.6	0.10
Bacillus megatherium	2.0	0.21
B. megatherium	2.4	0.24
Acetobacter suboxydans	2.7	0.54
B. megatherium	4.0	1.08
Cell-Free Preparations of		
A. aerogenes (Ø 0.6)		0.01
B. megatherium (Ø 2.0)		0.01
B. megatherium (Ø 4.0)		0.01

[a] Adapted from Langmuir.[11]

However, this is a biological parameter which cannot be subjected to control, so the most important resistance to the liquid-cell oxygen transfer is still the diffusion through the matrix of the aggregate.

C. Oxygen Concentration Profiles

When treating the transfer of oxygen from the bulk liquid into cell aggregates (cells immobilized in a matrix, pellets, or spontaneously formed microbial films) it is convenient to use mass balance equations and take the oxygen consumption into account as well. Considering the system to be one of heterogeneous catalysis the situation in a spherical aggregate of cells can be described by

$$\frac{d^2C}{dr^2} + \frac{2}{r} \cdot \frac{dC}{dr} = \frac{X_a}{D_a} \cdot \frac{q_{O_2}\,\max \cdot C}{K_m + C} \qquad (5)$$

where C = oxygen concentration in the aggregate at the depth r from the aggregate surface, X_a = cell density of the aggregate (weight/volume), D_a = diffusivity of oxygen in the aggregate, K_m = Michaelis constant for oxygen, and qO_2 max = the maximum specific respiration rate per weight of cells. Solution of this equation gives oxygen concentration profiles through the spherical particle of immobilized cell preparation. Biryukov and Tarasova[12] calculated oxygen and glucose gradients in microcolonies from this basis. A detailed theoretical treatment of the oxygen tension profiles in mycelial pellets is given by Kobayashi et al.[3]

However, experimental demonstrations of oxygen profiles in immobilized cell aggregates (such as those mentioned above) are rare. According to Equation 5, the oxygen profile depends on the diffusivity of oxygen through the matrix as well as on the oxygen consumption rate (which is the product of the cell density and the specific respiration rate). Situations may occur where all cells in an aggregate are active without oxygen limitation because of low oxygen consumption rates caused by other restrictions of the metabolism, while in another case most cells of the same preparation are oxygen

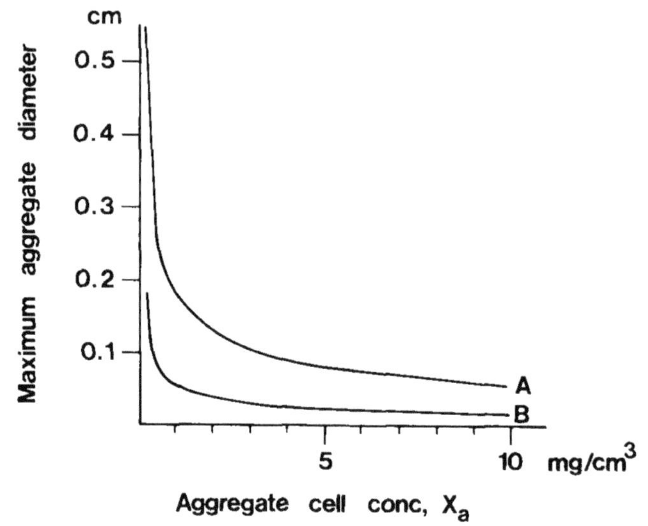

FIGURE 2. Maximum aggregate diameter that permits aerobic conditions throughout the aggregate as a function of the cell concentration (X_a) calculated according to data by Wuhrmann for two cases of specific oxygen uptake rate. A = qO_2 = 0.1 g/g·h. B: qO_2 = 1 g/g·h.

limited because of higher oxygen consumption of the cells in the peripheral layer in the aggregate. This has been observed by Bungay et al.[13]

Huang and Bungay[14] measured the oxygen profile through a 6 mm diameter *Aspergillus niger* pellet (dry weight 1.7 mg) and found that the oxygen tension began to decrease at 0.2 mm from the fixed pellet at a liquid flow rate of 28 cm/min. The profile inside the pellet depended on the glucose concentration, with the highest oxygen tension at the highest glucose concentration, a phenomenon that they ascribed glucose inhibition of respiration.

In order to elucidate the relation between maximum aggregate size for unlimited respiration and the oxygen consumption rate, the following formula derived from data by Wuhrman[15] can be used:

$$d_{crit} = \left[\frac{24\,D \cdot (C^* - C_{crit})}{q_{O_2} \cdot X_a}\right]^{-1/2} \qquad (6)$$

where d_{crit} = critical aggregate(sphere) diameter(cm), C^* = oxygen concentration at the aggregate surface (mg/cm³), C_{crit} = critical oxygen concentration for respiration, D = diffusivity of oxygen in the aggregate (cm²/sec), qO_2 = specific rate of respiration (mg/mg sec), and X_a = the cell density in the aggregate (mg/cm³). Figure 2 gives two curves that show how the critical aggregate size varies with the cell content at a relatively high respiration rate (1 g/g·h) and a low respiration rate (0.1 g/g·h), respectively. The diffusivity of oxygen was taken to be $5 \cdot 10^{-6}$ cm²/sec and C_{crit} was assumed to be 5% of an ambient medium oxygen concentration of 7 mg/ℓ. Similar results were obtained by Biryukov and Tarasova[12] who found that the critical concentration of oxygen surrounding a 0.4 mm diameter colony was 8 mg/ℓ.

Thus, it seems clear that if aerobic processes are to be performed with immobilized cells at volumetric rates corresponding to those of common fermentation processes, the aggregate size must be very small in order to ensure that the liquid-cell oxygen

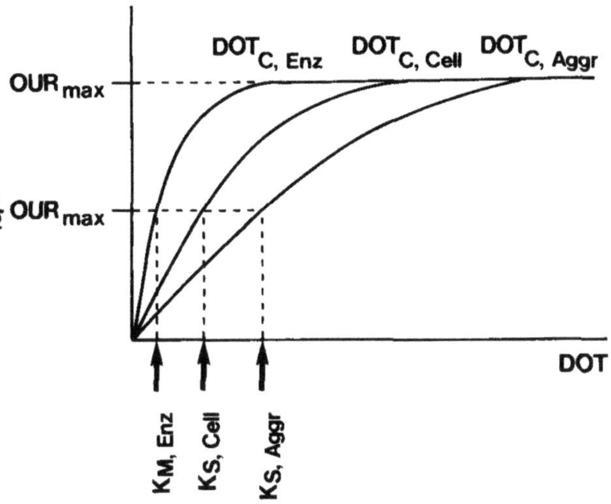

FIGURE 3. A schematic representation of the effect of the biocatalyst size on the Michaelis constant and the critical oxygen tension. OUR = oxygen uptake rate. Subscripts "enz", "cell" and "aggr" refer to cell free respiratory enzymes, whole cells and cells immobilized in aggregates, respectively. The true Michaelis constant of enzymes is designated K_M while the corresponding apparent constants for cells or cell aggregates are called K_s-values or saturation constants.

transfer is controlled only by molecular diffusion. The development of bioreactors that enable higher oxygen transfer rates to immobilized cells is therefore an important task for the biochemical engineers.

III. OXYGEN REQUIREMENT OF MICROORGANISMS

The relation between the oxygen uptake rate and the oxygen concentration may be described by the Michaelis-Menten model (Figure 3). The K_M-value, (which is the oxygen concentration that permits half the maximum reaction rate), is the most frequently used parameter for characterization of this relation, however, the more unprecise value for the critical concentration of oxygen is more important from the technical point of view, since it denotes the limit below which the rate of the process is limited by the oxygen concentration. In conventional fermentation processes efforts are made to ensure that the reaction rate is independent of oxygen by keeping the oxygen concentration arbitrarily above its critical concentration.

However, the literature values for oxygen dependence for microorganisms (k_s-values) are very difficult to evaluate since the experimental procedures are not described adequately. Critical oxygen concentrations varying from less than one percentage to over 100% (in air saturated medium) can be found. Table 3 gives a list of some reported data on K_s and critical concentrations. The reason for the scatter is that most investigators do not distinguish between "true" and "apparent" values. "True" values of K_s or critical concentration will be found only under conditions when the suspension of organisms is relatively diluted and well mixed, while apparent values are obtained in heterogeneous systems (Figure 3).

Interesting data on K_s-values which may approach "true" values have been published by Longmuir.[11] Table 2 is derived from data of Longmuir's work. The measurements were performed with dilute suspensions of organisms and show that K_s is a function of the cell size. Furthermore, the K_M of cell-free suspensions of the largest

Table 3
SOME LITERATURE DATA ON CRITICAL OXYGEN TENSIONS (DOT_{crit})

Organism	DOT_{crit}* (% air sat)	Ref.
Brevibacterium flavum	0.14	18
Torula utilis	1	19
Escherichia coli	2	16
Azotobacter vinlandii	4	20
Klebsiella aerogenes	6	17
Saccharomyces cerevisiae	8	25
Pseudomonas ovalis	14	26
Streptomyces niveus	42-140	21
Penicillium chrysogenum	6-75	22
	35	23
	160	24
Bacteria (general)	1.7-16	27
Bacteria and yeasts (general)	0.03-6	28
Molds (general)	6-21	27

* In some cases the DOT have been calculated from concentration data on the assumption that 100% air sat. equals 8 mg O_2/ℓ.

(B. megaterium) and the smallest *(A. aerogenes)* cells are almost equal in spite of the almost 100-fold larger K_S of the intact *B. megaterium*. A possible explanation of this is due to the formation of larger gradient around the larger cells. These cells have higher flows of oxygen per cell surface unit (if similar specific respiration rates are assumed, which was, however, not demonstrated). These gradients mean that the true oxygen concentration at the enzymatic level is lower than the oxygen concentration in the bulk liquid where the analysis is performed.

It is easy to imagine that aggregates of cells will behave as still larger cells with still larger gradients and therefore give rise to still larger observed K_s or critical oxygen concentrations.

While the "true" K_S and critical concentration are biological constants determined mainly by the properties of the cells, the corresponding apparent values are not of biological origin. They depend instead on the oxygen profile from the point of measurement of the oxygen concentration (mostly in the bulk liquid) to the cell. This profile is, as mentioned above, determined by the diffusivity of oxygen, the size of the aggregate and the volumetric oxygen consumption rate of the aggregate ($X_a \cdot qO_2$). In other words, the apparent values are reactor determined parameters. The highest values in Table 3 therefore only reflects high heterogeneity of the culture. Most of the oxygen is used for respiration[27] and data on critical oxygen concentration and K_S-values mainly concern this process. However, critical values of other reactions, those mediated by other oxidases and by oxygenases, may be higher since the respiratory oxidases seem to have unusually high affinity to oxygen. K_M-values of free glucose oxidases are generally in the range 0.5 m*M*, i.e., 200% air saturation.[29] Duarte and Lilly[30] observed that the rate of oxidation of cholesterol to cholestenone by cholesterol oxidase in *Nocardia rhodochrous* was linearily related to the dissolved oxygen tension up to 1 atm of oxygen. The reason might be a high true K_M value or a high apparent value caused by diffusional limitations.

IV. EXPERIMENTAL DATA ON OXYGEN TRANSFER RATES IN REACTORS WITH IMMOBILIZED CELLS

Several aerobic processes employing immobilized cells have been described in the literature, but quantitative data on the oxygen transfer is discussed only in very few cases. It is, however, possible to calculate the order of magnitude of oxygen consumption rates in several cases when the product formation kinetics and the stoichiometry of the process is known. Table 4 gives a list of oxygen consumption rates in processes with immobilized cells which have been derived in this manner. The conditions were different with respect to many parameters in these experiments and important data like aggregate size and cell density in the aggregate are not always reported, which makes a direct comparision meaningless. Still, the table reveals that aerobic processes with immobilized cells mostly deal with oxygen consumption rates that are much lower than corresponding values in processes with suspended cells. Presumably the reaction rate in most cases was limited by the oxygen transfer capacity of the system.

V. STRATEGIES TO INCREASE THE OXYGEN TRANSFER RATE TO IMMOBILIZED CELLS

A. Oxygen From External Sources

1. Recirculation of Aerated Medium

In fermentations involving suspended cells the main strategy for increasing the oxygen transfer rate is based on increased power input, viz., increased stirring rate, impeller design, and increased aeration rate. These strategies are not applicable to immobilized cells because of the mechanical limitations discussed below. Mere separation of the bulk liquid from the bioreactor to a traditionally aerated fermenter followed by recirculation of this air-saturated medium through the bioreactor will not permit extensive improvements. Calculations demonstrate that very high medium flow rate would be required. Take the example of a 1 ℓ reactor with 500 g gel containing 10 g cells. This is a typical cell load for a conventional reactor and therefore may be considered as a minimum load in a reactor with immobilized cells. Assume a specific glucose uptake rate of 1 g/g·h giving a total glucose uptake rate of 10 g/ℓ·h. This is a reasonable value to use since alginate immobilized *Clostridium acetobutylicum* have a total glucose uptake rate of 7 g/ℓ·h[38], as measured during butanol production. 10 g glucose/ℓ·h corresponds to 336 mmol oxygen/ℓ·h if the glucose is to be used for respiration or 28 mmol/ℓ·h if it is assumed that an oxidase reaction is needed to regenerate oxygen from hydrogen peroxide. Since air-saturated water dissolves approximately 0.25 mmol oxygen per liter the required liquid flow rate through the reactor would be 1500 ℓ/h, or 125 ℓ/h for the two cases (respiration and oxidase mediated substrate oxidation), respectively. The former case is far beyond the current possibilities. The calculation demonstrates that high productive aerobic processes cannot be achieved simply by aerating immobilized cells with air saturated medium.

Similar conclusions were drawn by Constantinides et al.[39] on the basis of data from glutamic acid synthesis by *Brevibacterium flavum* immobilized in collagen. They found that the maximum residence time was 6.6 sec in a column recycle reactor, i.e., a dilution rate of approximately 550.

2. Increased Oxygen Partial Pressure

One way to increase the oxygen transfer rate in the gas-liquid step as well as in the liquid-cell step is to increase the partial pressure of oxygen either by increasing the total air pressure or by increasing the concentration of oxygen in the gas phase. The latter method increases the gas-liquid transfer rate in proportion to the oxygen concen-

Table 4
OXYGEN CONSUMPTION RATES CALCULATED FROM PRODUCTIVITY OF IMMOBILIZED CELLS

Process	Reactor	Biocatalyst	Oxygen consumption (mol/ℓ·h)	Ref.
Sorbosone→2-keto-L-gulonic acid	Shake flask	*Pseudomonas syringae* in polyacrylamide-gel	4	31
Ribitol→erythropentulose	Shake flask	*Acetobacter suboxydans* glutaraldehyde-cross-linked	4	32
Threonine + glucose→isoleucine	Fluidized bed	Growing Serratia marcescens in carrageenan	10	33
Methionine→Keto- and -methiol-butyrate	Packed bed	*Trigonopsis variabilis* in Ca-alginate	10	34
Glycerol→dihydroxyacetone	Grounded beads	*Acetobacter* in polyacrylamide gel	11-36	35
Sucrose→gluconate + fructose	CSTR	Glucose oxidas and *S. cereviciae* in polyacryl amide-gel	26	36
Ethanol→acetic acid	Tower fermenter	*Acetobacter* sp. aggregated by hydrous TiO_2	83	37

tration according to Equation 3. When the air pressure (P) is increased, the expected oxygen transfer rate is proportional to $P^{1/3}$ if only the effect on C* and a are considered. The extended gas hold up, however, increases the OTR still more. According to Sato et al.,[40] the gas-liquid oxygen transfer rate at constant air mass flow increases in proportion to the 0.4 to 0.8th power of the air pressure in a stirred reactor.

Since no oxygen consumption is expected in the bulk liquid of a reactor with immobilized cells the equilibrium bulk liquid oxygen content will be proportional to the partial pressure of oxygen irrespective of which method is used to raise the partial pressure. This will then increase the liquid-cell oxygen transfer according to Equation 4.

Pure oxygen has been applied in several cases with immobilized cells. Pronounced effects on the conversion of glycerol to dihydroxyacetone by *Gluconobacter* have been reported by Makhotkina et al.[41] and by Holst et al.[42] From data on product formation kinetics published by Ghommidh et al.[43] it can be concluded that the oxygen consumption rate of acetic acid producing *Acetobacter* sp. adsorbed in porous ceramics increased from 77 mmol/ℓ·h to 173 mmol/ℓ·h when pure oxygen was substituted for air. From data by Couderc and Baratti[44] very high initial (20 min) oxygen consumption rates (75 mmol/ℓ·h) could be calculated during oxidation of methanol to formaldehyde by *Hansenula polymorpha* immobilized in polyacrylamide in a reactor purged with 1 VVM pure oxygen.

There are two major risks encountered with the application of pure oxygen. Carbon dioxide concentrations may reach inhibitory levels if respiration is involved and the oxygen itself may be toxic. The possibilities of increasing the oxygen transfer rate to immobilized cells by increasing the oxygen partial pressure of the gas phase are therefore restricted.

3. Oxygen Diffusivity and Particle Size

According to Equation 4 the rate of oxygen flow to the cells can be increased not only by increasing the bulk liquid oxygen concentration but also by increasing the diffusivity (D), or by reducing the mean distance of diffusion in the aggregate (x).

Karube and co-workers[45] compared the diffusivity of glucose in polyacrylamide, agar and collagen. They found agar to give the highest resistance and polyacrylamide to be the most permeable material to glucose, though the differences were relatively small. High cell contents (0.3 g cells wet weight/g gel) reduced the diffusivity of glucose in polyacrylamide by a factor of 33%.

One way to improve the oxygen supply to the cells would be to use matrices to which oxygen preferentially partition. In enzyme technology it has been possible to reduce the apparent K_M for a hydrophobic substrate[46] and the K_i for inhibitors[47] for a certain enzyme by making the immobilizing polymer more hydrophobic. The problem that arises here is that hydrophobic matrices normally are not very biocompatible. Even if such systems are developed, the positive effect would only be marginal.

A far more powerful technique would be to drastically change the particle size so that only small particles are used. This can of course be attained, but the problem then is the practical handling of the material. As shown in Figure 2, a particle size around 0.1 mm would be the upper limit in order to ensure a good oxygen supply for a dense immobilized cell population. Such small particles could very well be obtained and operational in, e.g., stirred batch or fluidized bed reactors, but probably not as effective in packed bed reactors. By using a matrix having a high specific density, the risk for a wash out in the former system can, however, be markedly reduced.

As is described in Chapter 7, Volume I, soluble immobilized cells in aqueous two-phase systems are encased in extremely small droplets of one of the two liquid phases which is dispersed in the other phase. In such a system very good mass transfer conditions are obtained. Use of thin layers of surface adsorbed cells may be considered as a special case of the reduction of the aggregate size. Points 1 to 3 so far have dealt with how to overcome the different oxygen diffusion restrictions in the water and matrix phases as illustrated in Figure 1. However oxygen may be carried by other vehicles than water.

4. Other Oxygen Carriers

In a project where glucose oxidase was used for analytical purposes, the enzyme was immobilized by a reversible biospecific binding to concanavalin A-Sepharose.[48] As in all such assays it was found that linearity was limited to 0 to 0.5 mM of glucose, because of oxygen deficiency. In those assays, the sample was administered in pulses. By introducing red blood cells containing hemoglobin it was possible to generate a pool of easily accessible oxygen within the support of the enzyme studied. In this case glucose oxidase and red blood cells were co-immobilized to the con A-Sepharose. On introducing a pulse of substrate the pO_2 within the matrix decreased concomitantly with the mobilization of the pool of oxygen stored in the hemoglobin packets of the cells (Figure 4). After the pulse had passed, pO_2 increased and the hemoglobin was reoxygenated.

When the substrate solution added to immobilized *Gluconobacter oxydans* was supplemented with hemoglobin it was found that conversion of glycerol to dihydroxyaceton was substantially improved[40] (Figure 5). However, the tendency of hemoglobin to spontaneously convert to methemoglobin may reduce the applicability of this method, at least for large-scale, long-term processes. In an effort to utilize the advantages obtained with hemoglobin as an oxygen carrier and avoid the disadvantages, synthetic carriers have been used instead.

Perfluorochemicals are organic compounds in which all hydrogen atoms have been replaced by fluorine atoms[49] (Figure 6). They are nonpolar, heat stable, and chemically inert. Many gases, for example oxygen and carbon dioxide, have high solubilities in perfluorochemicals. These properties make perfluorochemicals suitable as blood substitutes. Emulsion of perfluorochemicals have successfully been used in "blood-trans-

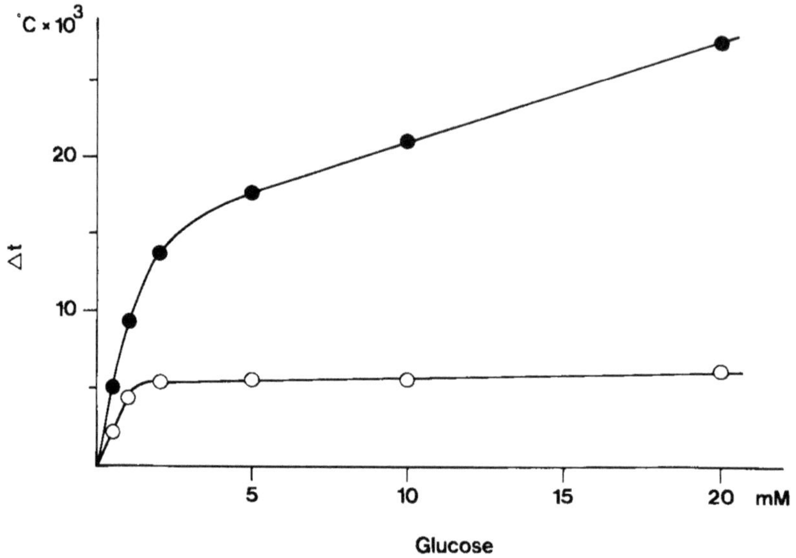

FIGURE 4. Analysis of glucose with an enzyme thermistor. Measured peak height (Δt°C) obtained from a glucose oxidase-concanavalin A-Sepharose column (○) or a glucose oxidase — red blood cell — concanavalin A-Sepharose column (●) as a function of the concentration of glucose dissolved in the perfusing buffer. (From Mattiasson, B. and Borrebaeck, C., *FEBS Lett.*, 85, 119, 1978. With permission.)

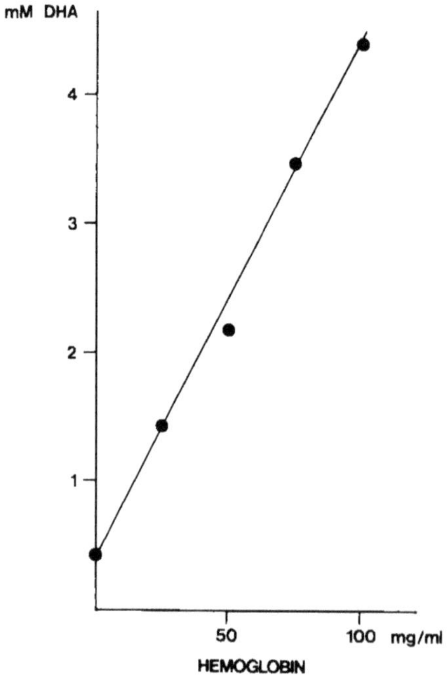

FIGURE 5. Production of dihydroxyacetone (μmol/min) by immobilized *Gluconobacter oxydans* as a function of the content of hemoglobin in the perfusing medium. (From Adlercreutz, P. and Mattiasson, B., *Eur. J. Appl. Microbiol. Biotechnol.*, in press.)

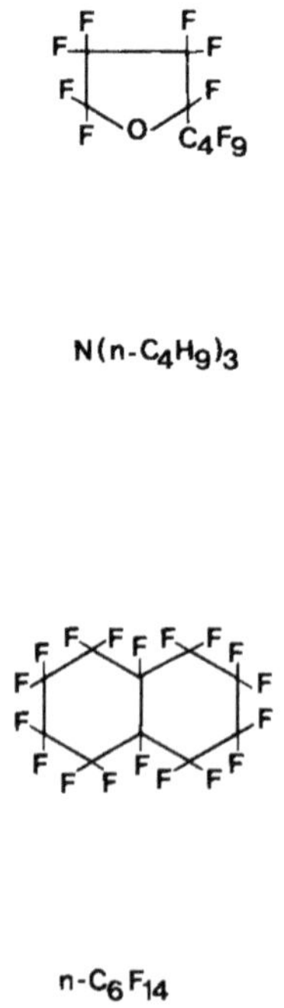

FIGURE 6. Chemical structure of some of the perfluorinated chemicals useful for oxygenation of immobilized cell-processes.

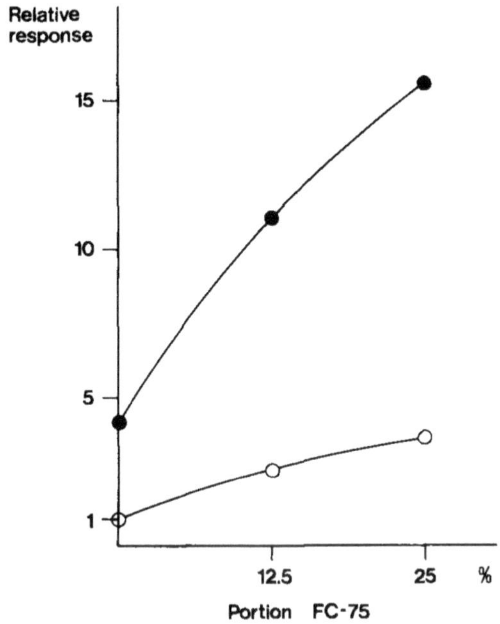

FIGURE 7. Test of the capacity of emulsions of the commercial mixture of perfluorinated chemicals, FC-75, to transport oxygen. Measurement of biologically available oxygen was carried out as a glucose analysis using an enzyme thermistor with glucose oxidase. (From Adlercreutz, P. and Mattiasson, B., *Eur. J. Appl. Microbiol. Biotechnol.*, in press.)

fusions'' to several animals,[50] and recently also to humans.[51] They have been found to be biocompatible. In an attempt to use such stable oxygen carriers instead of hemoglobin the same model reaction as that discussed above was studied. As seen from Figure 7 the use of 25% emulsions of the perfluorochemical FC-75, gives an increased amount of available dissolved oxygen (by a factor of 4). The same increase is also observed when the emulsion is equilibrated with pure oxygen. Long-term experiments showed no negative influence of the oxygen carrier on the behavior of the immobilized cells (Figure 8).

A different approach for increasing the partial pressure of oxygen in the medium was taken by Duarte and Lilly.[30] By using an organic solvent with a high capacity to dissolve oxygen instead of the normal buffer a marked increase in oxygen pressure was obtained in the solution surrounding the immobilized cells, thus ensuring a markedly improved supply of oxygen to the cells. This approach could be a useful one, but it must be stressed that the organic solvents used by Duarte and Lilly are generally very toxic to cells since they destroy the cell membranes.

Organic solvents can thus only be useful when one or a few enzymatic steps of the cellular metabolism is studied and only when the enzyme withstands the harsh treatment. No reports are so far available concerning the viability of the cells, but it seems reasonable to assume that they are killed by the treatment.

B. *In Situ* Oxygen Production

The mass transfer of oxygen as is illustrated in Figure 1 shows a rather complicated process, with many potential barriers. One way to improve the oxygen supply to immobilized cells would be to transport oxygen chemically bound in such a way that

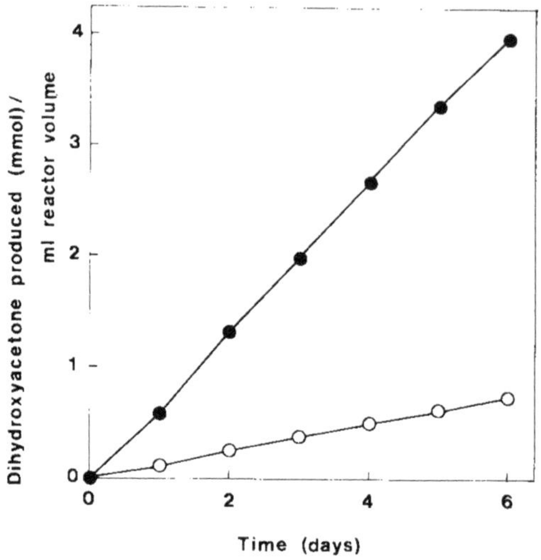

FIGURE 8. Long-term tests on the influence of perfluorinated chemicals on the catalytic capacity of the preparation of immobilized *Gluconobacter oxydans*. (From Adlercreutz, P. and Mattiasson, B., *Eur. J. Appl. Microbiol. Biotechnol.*, in press.)

allows it to be present in the external solution at higher concentrations than normal. In this way it could more easily penetrate into the immobilized cells, where another reaction has to take place in order to liberate the oxygen. Three different approaches to achieve this result will be discussed here.

1. Hydrogen Peroxide

In order to circumvent the difficulties in raising the oxygen tension in the aggregates of immobilized cells without applying increased gas liquid transfer or reduced aggregate size, Holst et al.[42] utilized the high catalase activity of *Gluconobacter oxydans* to convert hydrogen peroxide to oxygen *in situ*. The advantage of this system is that hydrogen peroxide can be administered with the medium. It then diffuses into the aggregates and is converted to oxygen by the influence of the catalase, in very close connection to subsequent utilization by the enzymes of interest. The problem of the transport of dissolved oxygen is then almost eliminated. On the other hand, the formation of oxygen from hydrogen peroxide must be carefully controlled. In this way a dramatic increase in the conversion of glycerol to dihydroxyacetone was achieved. However, the organism was impaired by the hydrogen peroxide at the concentration used and gradually lost its capacity after a few days. It was demonstrated to be important to keep the H_2O_2 concentration at a rather low level in the buffer, but still far above that of dissolved oxygen. Too high concentrations of hydrogen peroxide caused rapid cell death. This observation also led to the conclusion that hydrogen peroxide as an oxygen source must be used in connection with batch processes and not packed beds, since in the packed beds it is very difficult to utilize more than a fraction of the catalytic capacity unless there is a high excess of substrate. Since hydrogen peroxide had to be kept at a low concentration, batch processes were to be preferred (Figure 9). It should also be mentioned in this context that other catalysts than catalase may be used to generate oxygen from hydrogen peroxide. Manganese dioxide was used in early work by Lilly et al.[52] when they wanted to eliminate peroxide formed in an oxydase-catalyzed process. Many other investigators have found that charcoal and

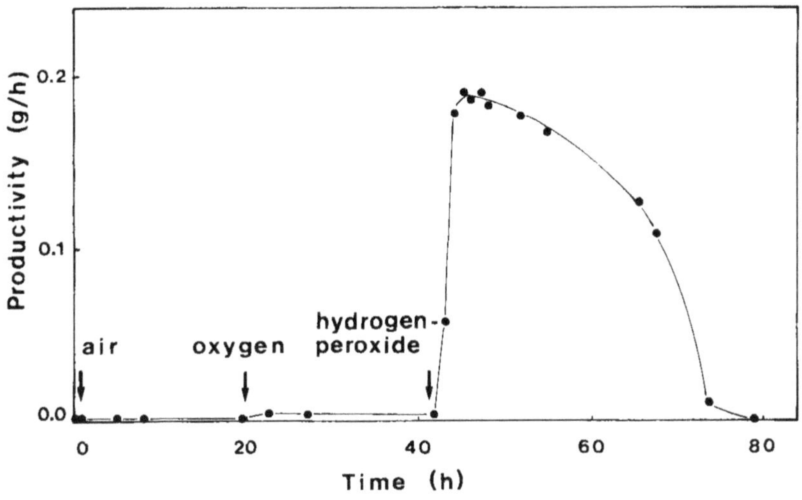

FIGURE 9. Continuous production of dihydroxyacetone from glycerol using *Gluconobacter oxydans* immobilized in Ca-alginate. The column (40 m*l* bed volume) was fed with 0.1 *M* glycerol in buffer, pH 5.0 (flow rate 1.5 m*l*/min). The source of oxygen was air oxygen and hydrogen peroxide (0.05 *M*), respectively, as indicated with arrows.

other inorganic materials, for example,[53] will decompose peroxides. However, in all of those processes some radicals are generated that may be toxic to the catalyst if not eliminated. For this purpose superoxide dismutase has been included as well.

Hydrogen peroxide has also been studied by Houtmeyers et al.[54] as a supplemental oxygen source in the activated sludge process. The damaging effects of hydrogen peroxide when used in the oxygenation of fermentation processes has been investigated by Ibrahim and Schlegel.[55] They observed low levels of hydrogen peroxide even when the catalase activity was in large excess. Several mechanisms of cell damage were discussed.

2. Photosynthetic Decomposition of Water as a Source to Generate Oxygen

Recently was reported by Adlercreutz and Mattiasson[56] on the use of immobilized algae for oxygen production. Photosynthesis is the most common way to generate oxygen in nature, and photolysis of water with the aid of immobilized *Chlorella pyranoidosa* was therefore utilized. *Gluconobacter oxydans* was used as a model organism to illustrate the effect of co-immobilization of an oxygen consumer with an oxygen producer (algae).[57] The cells were immobilized separately in alginate and then mixed in the column reactor or they were mixed before the immobilization procedure. A preparation containing only *Gluconobacter* was used for the control.

In preparing such symbiotic preparations several precautions have to be taken so that the cells will not negatively interfere with each others metabolism. For this reason it was first checked that the algae did not consume the bacterial substrate and also that no toxic exudates were produced by either cells. Figure 10 illustrates that the two columns containing both algae and bacteria were far superior to the one with bacteria only. As well, the preparation of co-immobilized cells was more efficient than the reactor containing the mixture of the two separately immobilized cell preparations. It would, at least in theory, be possible to generate efficient immobilized symbionts in the future.

The oxygen consumer, *Gluconobacter oxydans,* used in this model study is a favorable model since the oxygen consumption is not coupled to production in other gaseous products, e.g., carbon dioxide. When systems are used where the respiratory chain

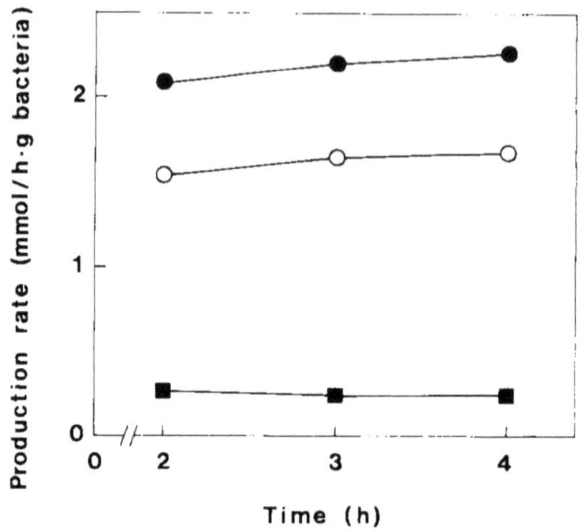

FIGURE 10. Productivity of *Gluconobacter oxydans* when immobilized alone (■) and packed in a column, packed together with a preparation of immobilized *Chlorella pyrenoidosa* (O) or when co-immobilized with the algae (●). All columns held the same content of bacteria and in the two with algae that amount was identical. The three columns were run under identical conditions. (From Adlercreutz, P., Holst, O., and Mattiasson, B., *Enzyme Microb. Technol.*, 4, 395, 1982.)

also is taking part in the reaction, gaseous products will cause problems of mass transfer since the carbon dioxide concentration will be enriched within the matrix. The problems with gas exchange in solid supports have very clearly been observed in anaerobic reactions, e.g., ethanol production by immobilized yeasts.[2] However, if such systems are set up where the oxygen generation is dependent upon photolysis of water one can also assume that the problem of carbon dioxide will be reduced in importance since the algae will recognize a favorable microenvironment within the co-immobilized gel preparation and convert carbon dioxide into carbohydrate.

3. Electrolytic Generation of Oxygen

Electrolytic decomposition of water to oxygen and hydrogen has been used in order to supply submerged fermentation with *Pseudomonas fluorescens* with oxygen.[58] However, no oxygen transfer measurements were reported and the authors noticed inhibitory effects which they ascribed to the possible formation of hypochlorite.

Oxygen transfer is also a problem in immobilized glucose oxidase and catalase used in enzyme electrodes for glucose analysis. Since this reaction depends on oxygen diffusion from the sample to the enzyme the measuring range is markedly restricted by the oxygen transfer process. In order to solve this problem, a glucose electrode has been designed where the enzymes are immoblized on a Pt-gauze which is connected as an anode in an electrolysis circuit.[59] The electrolysis current is controlled by the differential potential between the oxygen sensing part of the glucose electrode and a reference voltage (Figure 11). In this way the enzymatic reaction could be furnished with oxygen even at high glucose concentration (Figure 12) and in samples without any dissolved oxygen.[60]

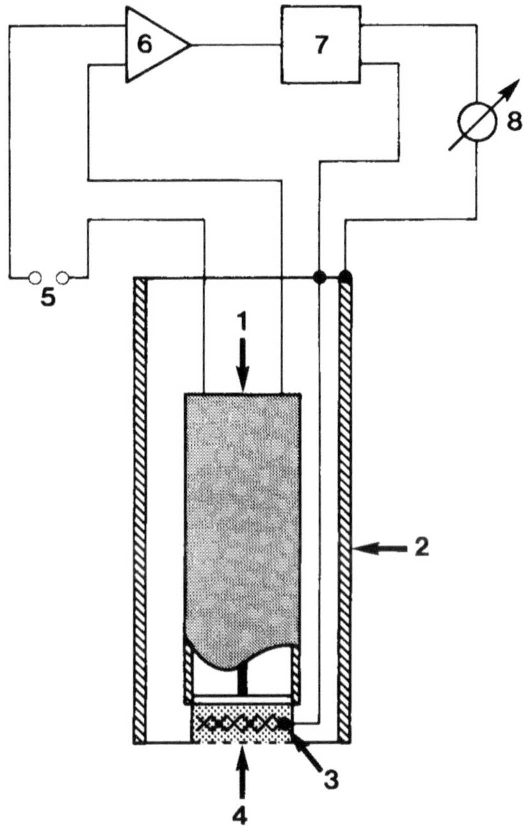

FIGURE 11. The principle of the oxygen stabilized enzyme electrode. 1. Oxygen electrode, 2. Stainless steelhouse, 3. Pt-gauze with immobilized enzyme, 4. Membrane, 5. Reference voltage, 6. Differential amplifier, 7. PID-controller for control of the electrolysis current. 8. μA-meter.

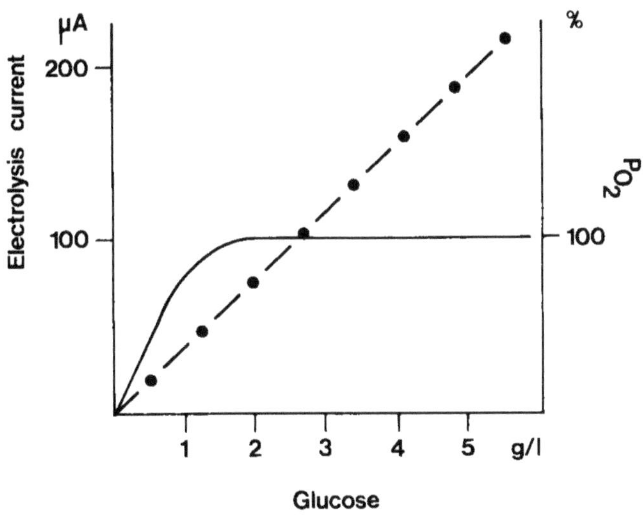

FIGURE 12. A comparison between the response of an ordinary oxygen diffusion dependant glucose electrode and the glucose electrode according to Figure 11 where the oxygen tension is stabilized by controlled electrolytic decomposition of water.

VI. CONCLUSION

Oxygen transfer is one of the bottlenecks in conventional fermentation technology and it has so far been almost totally overlooked with regards to immobilized cells. In the latter case with high cell densities it is realistic to assume that the oxygen transfer problems will become even more important. The present review covers the state of the art and presents some new concepts to improve supply to immobilized cells. It is, however, still far to go, and much research is needed within this area if oxygen consuming processes shall be fruitfully exploited using immobilized biocatalysts.

REFERENCES

1. Enfors, S. O. and Molin, G., The effect of different gases on the activity of microorganisms, in *Psychrotrophic Microorgansms in Spoilage and Pathogenicity,* Roberts, T. A., Hobbs, G., Christian, J. H. B., and Skovgaard, N., Eds., Academic Press, London, 1981, 335.
2. **Krouwel, P. G. and Kossen, N. W. F.,** Gas production by immobilized microorganisms: theoretical approach, *Biotechnol. Bioeng.,* 22, 681, 1980.
3. **Kobayashi, T., van Dedem, G., and Moo-Young, M.,** Oxygen transfer into mycelial pellets, *Biotechnol Bioeng.,* 15, 27, 1973.
4. **Cooper, C. M., Fernstrom, G. A., and Miller, S. A.,** Performance of agitated gas-liquid contactors, *Ind. Eng. Chem.,* 36, 504, 1944.
5. **Wise, W. S.,** The measurement of the aeration of culture media, *J. Gen. Microbiol.,* 5, 167, 1951.
6. **Bandyopadhyay, B., Humphrey, A. E., and Taguchi, H.,** Dynamic measurement of the volumetric oxygen transfer coefficient in fermentation systems, *Biotechnol. Bioeng.,* 9, 533, 1967.
7. **Tsao, G. T. and Lee, Y. H.,** Aeration, in *Annual Reports on Fermentation Processes,* Vol. 1, Perlman, D., Ed., Academic Press, New York, 1977, 115.
8. **Yoshida, T., Shimizu, T., Taguchi, H., and Teramoto, S.,** Studies on submerged culture of Basidiomycetes. III. The oxygen transfer within the pellets of *Lentinus edodes, J. Ferment. Technol.,* 45, 1119, 1967.
9. **Sada, E., Katoh, S., and Terashima, M.,** Enhancement of oxygen absorption by magnetite-containing beads of immobilized glucose oxidase, *Biotechnol. Bioeng.,* 23, 1037, 1981.
10. **Ngian, K. F. and Lin, S. H.,** Diffusion coefficient of oxygen in microbiol aggregates, *Biotechnol. Bioeng.,* 18, 1623, 1976.
11. **Longmuir, I. S.,** Respiration rate of bacteria as a function of oxygen concentration, *Biochem.,* 57, 81, 1954.
12. **Biryukov, V. V. and Tarasova, S. S.,** Permeability criterion of microcolonies, *Biotechnol. Bioeng. Symp.,* 4, 573, 1973.
13. **Bungay, H. R., Whalen W. J., and Sanders, W. M.,** Microprobe techniques for determining diffusivities and respiration rates in microbial slime systems, *Biotechnol. Bioeng.,* 11, 765, 1969.
14. **Huang, M. Y. and Bungay, H. R.,** Microprobe measurement of oxygen concentrations in mycelial pellets, *Biotechnol. Bioeng.,* 15, 1193, 1973.
15. **Wuhrmann, K.,** Microbial aspects of water pollution control, *Adv. Appl. Microbiol.,* 6, 119, 1964.
16. **Kempner, W.,** The effect of oxygen tension on cellular metabolism, *J. Cellular Comp. Physiol.,* 10, 339, 1937.
17. **Harrison, D. E. F. and Pirt, S. J.,** The influence of dissolved oxygen concentration on the respiration and glucose metabolism of *Klebsiella aerogenes* during growth, *J. Gen. Microbiol.,* 46, 193, 1967.
18. **Hirose, Y., Akashi, K., and Shibai, H.,** Effect of oxygen tension on the microbial production of amino acids, in *Abstracts of VIth Int. Fermentation Symp.,* National Research Council, Ottawa, Canada 1980, 36.
19. **Johnson, M. J.,** Aerobic growth at low oxygen concentrations, *J. Bacterial.,* 94, 101, 1967.
20. **Fife, J. M.,** The effect of different oxygen concentrations on the rate of respiration of *Azotobacter* in relation to the energy involved in nitrogen fixation and assimilation, *J. Agr. Res.,* 66, 421, 1943.
21. **Fewkes, R. C. J. and Wang, D. I. C.,** Physical parameters governing the behavior of non-newtonian mycelial antibiotic fermentations, in Abstract of Papers, Fifth Int. Fermentation Symp., Berlin, 1976.
22. **Phillips, D. H. and Johnson, M. J.,** Aeration in fermentations, *J. Biochem. Microbiol. Technol. Eng.,* 3, 277, 1961.

23. Fish, N. M., Vardar, F., and Lilly, M. D., Effects of dissolved gas concentrations on microbial product formation, in Abstracts of Communications, 2nd European Congress of Biotechnology, Eastbourne, 1981.
24. Giona, A. R., De Santis, R., Marrelli, L., and Toro, L., The influence of oxygen concentration and of specific rate of growth on the kinetics of penicillin production, *Biotechnol. Bioeng.*, 18, 493, 1976.
25. Brown, D. E., Aeration in submerged culture, in *Methods in Microbiology*, Norris, J. R. and Ribbons, D. W., Eds., Academic Press, London 1970, 125.
26. Bennet, G. F. and Kempe, L. L., Oxygen transfer mechanisms in the gluconic acid fermentation by *Pseudomonas ovalis*, *Biotechnol. Bioeng.*, 6, 347, 1964.
27. Mukhypadhyay, S. N. and Ghose, T. K., Oxygen participation in fermentation. I. Oxygen microorganism interactions, *Process Biochem.*, 11, 19, 1976.
28. Harrison, D. E. F., MacLennan, D. G., and Pirt, S. J., Responses of bacteria to dissolved oxygen tension, in *Fermentation Advances*, Perlman, D., Ed., Academic Press, London, 1969, 117.
29. Linek, V., Beneš, P., Sinkule, J., Holoček, O. and Malý, V., Oxidation of D-glucose in the presence of glucose oxidase and catalase, *Biotechnol. Bioeng.*, 22, 2515, 1980.
30. Duarte, J. M. C. and Lilly, M. D., The use of free and immobilized cells in the presence of organic solvents: the oxidation of cholesterol by *Nocardia rhodochrous*, *Enz. Eng.*, 5, 363, 1980.
31. Martin, C. K. A. and Perlman, D., Conversion of L-sorbose to 2-keto-L-gulonic acid by mixtures of immobilized cells of *Gluconobacter melanogenus* IF03293 and *Pseudomonas* species, *Eur. J. Appl. Microbiol.*, 3, 91, 1976.
32. Schnarr, G. W., Szarek, W. A., and Jones, J. K. N., Preparation and activity of immobilized *Acetobacter suboxydans* cells, *Appl. Environ. Microbiol.*, 33, 732, 1977.
33. Wada, M., Uchida, T., Kato, J., and Chibata, I., Continuous production of L-isoleucin using immobilized growing *Serratia marcescens* cells, *Biotechnol. Bioeng.*, 22, 1175, 1980.
34. Brodelius, P., Hägerdal, B., and Mosbach, K., Immobilized whole cells of the yeast *Trigonopsis variabilis* containing D-amino acid oxidase for production of α-keto acids, *Enz. Eng.*, 5, 383, 1980.
35. Nabe, K., Izuo, N., Yamada, S., and Chibata, I., Conversion of glycerol to dihydroxyacetone by immobilized whole cells of *Acetobacter xylinum*, *Appl. Environ. Microbiol.*, 38, 1056, 1979.
36. D'Souza, S. F. and Nadkarni, G. B., Continuous conversion of sucrose to fructose and gluconic acid by immobilized yeast cell multienzyme complex, *Biotechnol, Bioeng.*, 22, 2179, 1980.
37. Kennedy, J. F., Humphreys, J. D., and Barker, S. A., Application of living immobilized cells to the acceleration of the continuous conversions of ethanol (wort) to acetic acid (vinegar)-hydrous titanium (IV) oxide-immobilized *Acetobacter* species, *Enz. Microb. Technol.*, 2, 209, 1980.
38. Häggström, L., Immobilizied cells of *Clostridium acetobutylicum* for butanol production, in *Advances in Biotechnology*, Vol. 2, Moo-Young, M. and Robinson, C. W., Eds., Pergamon Press, Toronto, 1981, 79.
39. Constantinides, A., Bhatia, D., and Vieth, W. R., Immobilization of *Brevibacterium flavum* cells on collagen for the production of glutamic acid in a recycle reactor, *Biotechnol. Bioeng.*, 23, 899, 1981.
40. Sato, S., Mukataka, S., Kataoka, H., and Takahashi, J., Oxygen absorption rate in an aerated stirred tank under increasing pressure, *J. Ferment. Technol.*, 59, 221, 1981.
41. Makhotkina, T. A., Pomortseva, N. V., Lomova, I. E., and Nikolaev, P. I., Glycerol transformation into dihydroxyacetone by polyacrylamide gel immobilized cells of *Gluconobacter oxydans*, *Prikladnaja Biochimija i Mikrobiologija*, 17, 102, 1981.
42. Holst, O., Enfors, S. O., and Mattiasson, B., Oxygenation of immobilized cells using hydrogen peroxide: a model study of *Gluconobacter oxydans* converting glycerol to dihydroxyacetone, *Eur. J. Appl. Microbiol.*, 14, 64, 1982.
43. Ghommidh, C., Navarro, J. M., and Durand, G., Acetic acid production by immobilized *Acetobacter* cells, *Biotechnol. Lett.*, 3, 93, 1981.
44. Couderc, R. and Baratti, J., Immobilized yeast cells with methanol oxidase activity: preparation and enzymatic properties, *Biotechnol. Bioeng.*, 22, 1115, 1980.
45. Matsunaga, T., Karube, I., and Suzuki, S., Some observations on immobilized hydrogen producing bacteria: behaviour of hydrogen in gel membranes, *Biotechnol. Bioeng.*, 22, 2607, 1980.
46. Reese, E. T. and Mandels, M., Enzyme action on partition chromatographic columns, *J. Am. Chem. Soc.*, 80, 4625, 1958.
47. Filippusson, H. and Hornby, W. E., The preparation and properties of yeast β-fructofuranosidase chemically attached to polystyrene, *Biochem. J.*, 120, 215, 1970.
48. Mattiasson, B. and Borrebaeck, C., An analytical flow system based on reversible immobilization of enzymes and whole cells utilizing specific lectin-glucoprotein interactions, *FEBS Lett.*, 85, 119, 1978.
49. Adlercreutz, P. and Mattiasson, B., *Eur. J. App. Microbiol. Biotechnol.*, in press.
50. Reiss, J. G. and LeBlanc, M., Perfluor-Verbindungen als Blutersatzmittel, *Angew. Chem.*, 90, 654, 1978.

51. **Schnoy, N., Pfannkuch, F., and Beisbarth, H.,** Fluorocarbone als Erythrozytenersatz, *Anaesthesist,* 28, 503, 1979.
52. **Carter, R. S., Prenosil, J. E., and Bourne, J. R.,** Stability studies on the immobilized glucose oxidase/catalase system, *Enz. Eng.,* 5, 321, 1981.
53. **Cho, Y. K. and Bailey, J. E.,** Enzyme immobilization on activated carbon: alleviation of enzyme deactivation by hydrogen peroxide, *Biotechnol. Bioeng.,* 19, 769, 1977.
54. **Houtmeyers, J., Poffé, R. and Verchtert, H.,** Hydrogen peroxide as a supplemental oxygen source for activated sludge: microbiological investigations, *Eur. J. Appl. Microbiol.,* 4, 295, 1977.
55. **Ibrahim, M. and Schlegel, H. G.,** Oxygen supply to bacterial suspensions of high cell densities by hydrogen peroxide, *Biotechnol. Bioeng.,* 22, 1877, 1980.
56. **Adlercreutz, P. and Mattiasson, B.,** Oxygen supply to immobilized cells. I. Oxygen production by immobilized *Chlorella pyrenoidosa, Enzyme Microb. Technol.,* 4, 332, 1982.
57. **Adlercreutz, P., Holst, O., and Mattiasson, B.,** Oxygen supply to immobilized cells. II. Studies on a co-immobilized algae-bacteria preparation with *in situ* oxygen generations, *Enzyme Microb. Technol.,* 4, 395, 1982.
58. **Sadoff, H. L., Halvoron, H. O., and Finn, R. K.,** Electrolysis as a means of aerating submerged cultures of microorganisms, *Appl. Microbiol.,* 4, 164, 1956.
59. **Enfors, S. O.,** Oxygen-stabilized enzyme electrode for D-glucose analysis in fermentation broths, *Enz. Microb. Technol.,* 3, 29, 1981.
60. **Enfors, S. O.,** A glucose electrode for fermentation control, *Appl. Biochem. Biotechnol.,* 7, 1982.

Chapter 4

FERMENTATIONS BY IMMOBILIZED STRICT ANAEROBES

Lena Häggström

TABLE OF CONTENTS

I. INTRODUCTION

"Fermentations can be defined as those biological processes that occur in the dark and that do not involve respiratory chains with oxygen or nitrate as electron acceptors".[1] Bacteria carrying out fermentations are anaerobes but in order to make a meaningful distinction between immobilized anaerobic cells and other immobilized cells it is necessary to restrict the term anaerobic to those organisms that really are sensitive towards molecular oxygen. In fact, there is no sharp borderline between anaerobes and aerobes but the whole scale of organisms tolerating increasing levels of oxygen are present. For the ease of discussion the following grouping can be made: (1) strict anaerobes which do not use molecular oxygen and also are very sensitive to the toxic effects of oxygen; (2) aerotolerant anaerobes also do not use oxygen and carry out a fermentative metabolism but they tolerate and can even grow in the presence of oxygen; (3) facultative anaerobes can shift between fermentative and respiratory metabolism; (4) strict aerobes which have an absolute requirement for oxygen.

Regarding the use of immobilized cells in a process there are not any distinct differences between the four categories. Most of the general advantages and disadvantages are valid for all types of cells. But one fact that distinguishes anaerobes from other organisms is that they must be protected from oxygen during the immobilization procedure. This applies only to the strict anaerobes. Once immobilized and placed in a reactor the need for anaerobic conditions should not impose any further problems if general anaerobic techniques are used. As a matter of fact, the immobilization may offer a protection against oxygen or high oxidation-reduction potential of the medium, which reduces the difficulties in keeping low E_h or oxygen tension for the organisms. Using aerotolerant anaerobes or facultatively anaerobic organisms in an immobilized process it might not even be necessary to exclude oxygen from any stage of the process.

Although the exclusion of oxygen might create practical problems during the immobilization procedure the fact that oxygen is not needed is also a great advantage. The problem with oxygen transfer to immobilized cells has not yet been solved satisfactorily.[2] On the contrary, gaseous products like carbon dioxide, hydrogen, and methane are often produced abundantly and the problem of gas desorption is sometimes substituted for the oxygen transfer problem.

The present literature on immobilized strict anaerobes is rather limited probably due to difficulties in handling these organisms. The following review will be confined to such systems, i.e., mainly methane producing bacteria and clostridia. The main purpose of these processes is to produce the primary products of the energy metabolism in the organism. These products are methane, hydrogen, fatty acids, and solvents. A common property of these systems is that they constitute microbiological energy conversion processes in the sense that the products technically are more fitted as fuels or chemical feedstocks than are the original substrates.

II. METHANE

The capability of microorganisms to produce methane has long been exploited in the treatment of solid wastes and concentrated waste waters. The most widely used process is the digestion of activated sludge. With regard to energy requirements anaerobic treatment of waste waters is becoming an attractive alternative to aerobic processes like the activated sludge process. However, with present technology it is hardly possible to apply anaerobic digestion to dilute waste waters like municipal sewage. Anaerobic metabolism yields very little biomass and long times would be needed to build up a stable microflora. If no means of biomass retention is introduced long hydraulic retention times must be applied for a complete conversion of the waste material. Such a system would also be very sensitive to disturbances.

Experiences from aerobic waste water treatment show that the efficiency of a process can be improved by increasing the biomass concentration. This is utilized in the activated sludge process where spontaneously produced bacterial polysaccharides form a gelatineous matrix in which other organisms are entrapped. The beneficial effects of the rotary disk contactors and trickling filter processes where a bacterial film develops and is maintained on solid surfaces are also well established. It thus seems reasonable to assume that the immobilization of a methanogenic population would lead to an increased overall efficiency of the anaerobic process.

In fact, recent research in the field tend to develop in this direction. Dilute waste water (30 to 110 ppm COD) has been treated in a packed bed reactor on a pilot plant scale.[3] The packing material used was ceramic Raschig rings to which the organisms adsorb. After start up 50% reduction of the BOD was achieved at a residence time of 6 to 35 hr with loading rates of 24 to 192 mg COD/ℓ reactor-volume · day. The methane production rate was, however, low at these conditions. It has also been reported that adsorption of organisms to sand particles used in a fluidized bed system greatly enhances the gas production rate and decreases the treatment time.[4] At an inflowing COD concentration of 9100 mg/ℓ and at a loading rate of 22.2 g COD/ℓ · day the methane production rate achieved was 3.3 ℓ/ℓ · day at a hydraulic retention time of 7.3 hr. These investigations have one thing in common. They utilize the ability of organisms to spontaneously immobilize by adsorption to the support material. Recently developed techniques for immobilization enable us though to immobilize all kinds of organisms irrespective of their ability to spontaneously adsorb to support materials.

With the aim of developing a continuous methane production system from waste waters Karube et al.[5] have immobilized a mixed methanogenic culture of unknown composition. Different immobilization materials were tried (agar, polyacrylamide, and a collagen membrane) of which agar was found to give the best results when incubated in waste water from an alcohol factory. Optimum agar concentration, optimum biomass concentration, optimum pH, and temperature were further established. When the same amount of free and immobilized cells were incubated in the waste water, the free cells had a higher initial rate of methane production but the total amount of gas produced was higher for the immobilized cells. In a mixed culture methane is mainly formed from the intermediates acetate or hydrogen and carbon dioxide produced by fermentative and acetogenic bacteria. Due to the immobilization, these products are formed very close to the methane producing bacteria and thereby more readily available. This was thought to apply especially to the gases. Similar mechanisms have been demonstrated for co-immobilized enzymes.[6] On the other hand it can be expected that locally high concentrations of carbon dioxide will build up inside the gel matrix. High partial pressures of carbon dioxide have been shown to inhibit methanogenic bacteria.[7]

It was also concluded that free cells of this population were more sensitive to oxygen than were immobilized cells. The culture was incubated in air for a 6-hr period in repeated batches. After the second run no methane activity could be detected for the free cells whereas the immobilized cells showed no inactivation at all during four successive runs. Obviously the presence of facultative organisms at a close distance protected the strict anaerobes from the harmful effects of oxygen.

Continuous methane production was simulated by incubating the immobilized population (1 g agar gel containing 0.02 g wet cells) in waste water (5 mℓ) in successive 24-hr batches for a total of 90 days under aerobic conditions. In this case the authors got an initial increase in methane production due to growth of organisms in the agar gel. After 25 days a steady state was reached with the methane production rate 4.5 μmol/cm^{3}gel · hr (Figure 1) or 4.5 mmol/ℓ gel · hr. In order to make this figure comparable to others, let us assume that 1 ℓ of gel is equivalent to 1 ℓ reactor volume. At

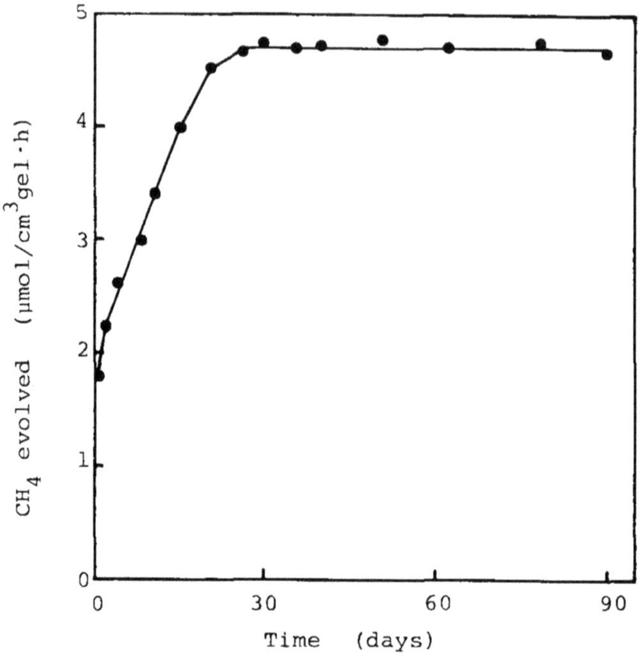

FIGURE 1. Methane production in consecutive batches by an agar immobilized population. (From Karube, I., et al., *Biotechnol. Bioeng.*, 22, 847, 1980. With permission.)

25°C the rate of methane production would then be 2.7 ℓ/ℓ·day. In a conventional sludge digestor the gas production is in the range of 0.5-1ℓ/ℓ·day. The waste water (BOD 1000 ppm) contained glucose (0.36 g/ℓ), sucrose (0.54 g/ℓ), and small amounts of formic, acetic, and propionic acids. The sugars and formic acid seemed to be completely utilized after 18 hr of incubation but the other organic acids remained in the medium.

Scherer et al.[8] immobilized a pure culture of the methanogenic organism *Methanosarcina barkeri* in a calcium alginate gel. All procedures in this investigation were performed at strict anaerobic conditions since this organism is very sensitive to oxygen having a half-life time of 4 min in air. The authors found that the immobilization did not protect the organism from oxygen. This is rather natural since in this case no oxygen is consumed by facultative organisms and the gel itself does not exclude oxygen but only delays its transport. The experiments were performed in a defined minimal medium with methanol as the sole carbon source. No diffusion limitations seemed to occur in this system as judged from the methanol conversion rates. Gel beads of different sizes gave the same conversion rates as did free cells initially. It was shown that larger beads preserved the microbial activity better than small beads, that entrapped cells were active longer than free cells, and that incubation of the entrapped cells in a growth medium maintained the activity still better. In a buffer solution the half-life of the system was 4 days. The authors have calculated the methane production rate of their system to be 15.8 ℓ/ℓ reactor-volume·day.

All the examples cited above show that definite advantages can be achieved in the treatment of waste water or in the production of methane if some kind of immobilized cells are used. However, agar and alginate immobilized cells are very remote from a practical application in this area. This may not have been the intention either. Nevertheless, immobilization in such materials may provide a useful basis for the study of the complex system that methane production constitutes.

III. HYDROGEN

An alternative to methane production is the production of hydrogen from wastes. Saccharolytic clostridia produce this compound from carbohydrate sources. One of the most potent hydrogen producers is *Clostridium butyricum* from which 2.35 moles of hydrogen per mole of glucose consumed have been obtained.[9] Simultaneously 2 moles of carbon dioxide are produced and the rest of the carbon skeleton is converted to butyric and acetic acids. Under unfavorable conditions, e.g., iron deficiency or oxygen exposure, the ferredoxin-hydrogenase system is not capable of carrying out the required reactions. The precursor pyruvate is instead converted to lactic and formic acids.[10]

In a series of articles by the Japanese group Karube, Matsunaga, Suzuki, and others the anaerobic bacterium *Clostridium butyricum* (IFO 3847) has been immobilized in various carriers, the purpose of the work being to utilize the hydrogen produced in a biochemical fuel cell for production of electrical energy. In the first paper basic properties were investigated when the organism was immobilized in a polyacrylamide gel.[11] The bacterium seemed to be rather sensitive to the monomers used for gel preparation. A total concentration of 10% was chosen in order to prepare a stable gel, although this amount of monomers clearly reduced the activity of the cells. According to the authors the immobilization protects the organism from oxygen. Free cells were very sensitive to oxygen but in the immobilized state they were active for at least 5 days when incubated aerobically (Figure 2). Continuous hydrogen production was simulated by means of repeated 24-hr batches for a total of 20 days. The immobilized cells were then incubated anaerobically in a phosphate buffer supplemented with glucose. The initial rate of hydrogen production was 0.525 mmol/day·g wet cells but decreased slowly during the experimental period.

In the following investigation *Clostridium butyricum* was immobilized in a polyacrylamide gel that was prepared from 3.6% of monomers.[10] A 0.1 cm thick gel layer containing totally 0.4 g wet cells covered a platinum cylinder which was used as anode in a fuel cell. The anode was placed in a separate chamber containing phosphate buffer and glucose. As the cathode was used a carbon electrode was in contact with a phosphate buffer and air. The two chambers were connected by a salt bridge. The products formed by the immobilized cells were formic acid, acetic acid, lactic acid, and hydrogen with a yield coefficient of respectively 0.2, 0.6, 0.15, and 0.6 mol/mol glucose. Butyric acid was probably also produced. Of these products hydrogen and formic acid were electroactive, thus creating a current in the fuel cell when produced in the anode chamber. When the authors used the fuel cell as described above (the anolyte was changed every third day) a constant current of 1.1 to 1.2 mA (no resistance between the electrodes) was obtained for 15 days. Later work showed, however, that the diffusibility of hydrogen towards the electrode surface was rather limited in this construction.[12]

A slight modification of this fuel cell was then utilized as a BOD sensor.[13] A standard solution containing glucose and glutamic acid was used to calibrate the fuel cell. A linear relationship between the steady-state current and the BOD was obtained up to BOD values of 300 ppm. Then waste waters from different sources (slaughter house, fermentation waste water from the laboratory, alcohol factory, and food factory) were diluted and analyzed in the fuel cell. The data obtained were compared with BOD_5 values determined by a standard method. A relatively good correlation between the two methods were found. The bio fuel cells deviation being within ± 10% from the BOD_5 test. As the authors also point out this method can be used only for certain kinds of waste water. Normal BOD estimation are based on a complex microflora being able to attack various kinds of organic compounds. When a pure culture is used

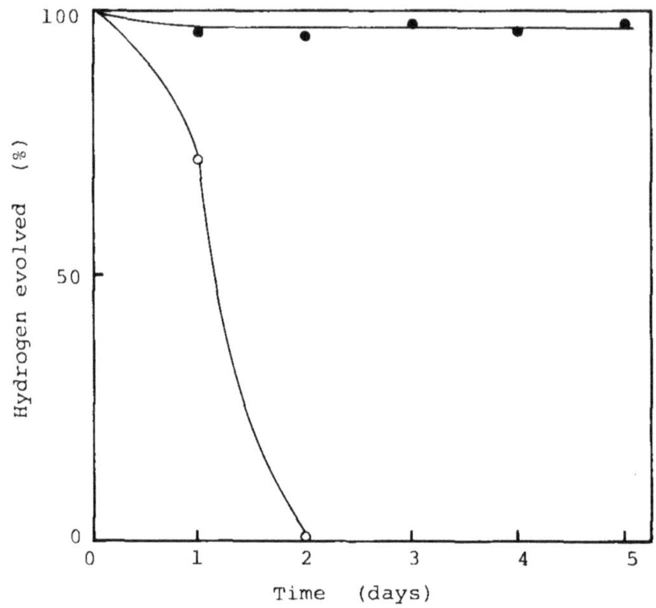

FIGURE 2. Hydrogen evolution by free (—o—) and polyacrylamide immobilized (—•—) cells of *Clostridium butyricum* at aerobic conditions. (From Karube, I., et al., *Biochim. Biophys. Acta*, 444, 338, 1976. With permission.)

many substrates may not be attacked and others may be inhibitory. Moreover, different substrates will give different product formation patterns and since the products have different electrochemical activity this will undoubedly influence the results.

The system was further developed in order to obtain electrical energy from waste waters.[14] The idea was now to produce hydrogen gas in a separate reactor containing the immobilized cells. The waste water was then transported to the anode chamber in the fuel cell where dissolved hydrogen and other electroactive products reacted and produced electrical energy. The anode and cathode chambers were also separated by an ion exchange membrane. This time *Clostridium butyricum* was immobilized in an agar gel because it preserved the microbial activity better than the polyacrylamide gel. Waste waters from an alcohol factory, a food factory, a slaughter house, and a glucose solution with the same intial BOD (660 ppm) were then compared with respect to hydrogen production. A rather wide variation was found ranging from 8 μmol H_2/h·g wet cells for the slaughter house water to 72 μmol H_2/h·g wet cells for the alcohol factory waste water. It is difficult to understand how these results can be compatible with the use of this system as a BOD sensor as described earlier.[13] However, in the final experimental set up, waste water from an alcohol factory (BOD 660 ppm) was continuously pumped (5 mℓ/min) through a packed bed reactor (90 mℓ volume) with 40 g of agar gel in small blocks containing a total of 4 g wet cells. The substrate was then transferred to the fuel cell, and finally to a stirred reactor for aerobic BOD reduction by a mixed culture (30 g cells in 300 g agar blocks, reactor volume approximately 1 ℓ). In a 20-day long run a steady-state current of 13 to 15 mA was obtained from the fuel cell and the waste water leaving the reactor had a BOD of 50 ppm. During these conditions the hydrogen production was reported to be 8 μmol/mℓ waste water, which can be recalculated to 0.6 mmol/h·g wet cells.

In an attempt to improve the efficiency of the biochemical fuel cell Suzuki et al.[15] have recently used a stirred reactor containing the immobilized cells and a device for

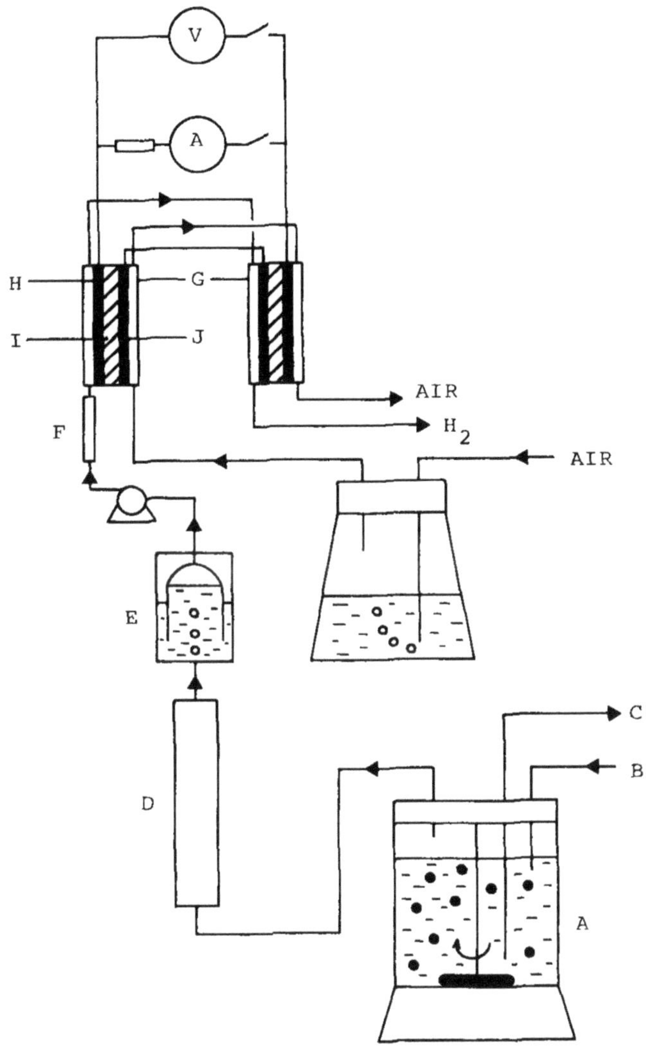

FIGURE 3. System for hydrogen production by agar immobilized *Clostridium butyricum* and the biochemical fuel cell arrangement. (A) Fermenter with immobilized cells. (B) Medium inlet. (C) Medium outlet. (D) Column with soda lime. (E) Hydrogen reservoir. (F) Flow meter. (G) Fuel cells. (H) Anode. (I) Nylon filter with potassium hydroxide. (J) Cathode. (From Suzuki, S., et al., *Biochimie*, 62, 353, 1980. With permission.)

separating the hydrogen from the liquid before it is transferred to the fuel cell (Figure 3).[15] *Clostridium butyricum* was immobilized in an agar gel, cut into small blocks (5 mm^3). One kilogram of gel containing 100 g of bacteria was placed in a laboratory fermenter with 5 *ℓ* of waste water from the alcohol factory. For maximum hydrogen production it was necessary to use condensed waste water (BOD 21000 ppm). Increased stirring rate increased the hydrogen production rate but the agar gel was gradually crushed by the agitation. The smaller gel particles might thus be responsible for the higher hydrogen evolution due to less diffusion resistance. Continuous hydrogen production was tested in repeated batches. After 4 days the hydrogen evolution decreased due to the disintegration of the gel. This problem was solved by the addition of 200 g of fresh gel at 5-day intervals. With this procedure the authors claim that it is possible

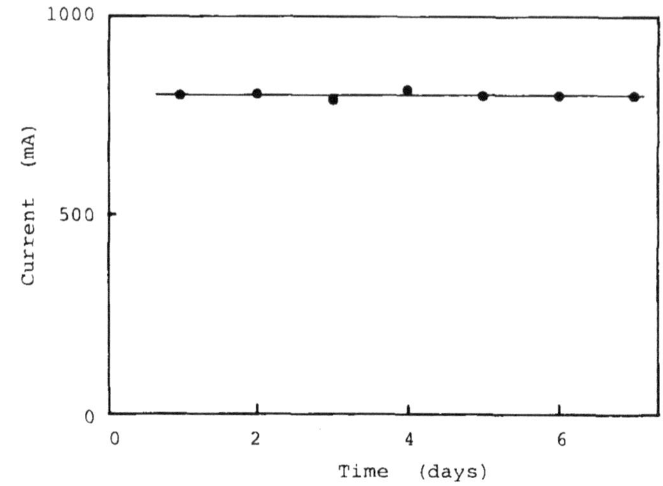

FIGURE 4. Current produced in the biochemical fuel cell system. (From Karube, I., et. al., *Ann. N.Y. Acad. Sci.*, 369, 91, 1981. With permission.)

to obtain continuous hydrogen production for more than 20 days at a rate of 6 m*ℓ* H_2/min·kg gel, which equals 0.14 mmol H_2/h·g wet cells (1 atm, 37°C). The fuel cell was also improved. The anode (platinum black-nickel mesh) and the cathode (palladium black-nickel mesh) was separated by a nylon filter containing potassium hydroxide. Two such cells were connected in a series. Pure hydrogen gas (6 m*ℓ*/min) was then transferred from the fermenter to the anode chambers and air to the cathode chambers (Figure 3). A current of 500 to 550 mA, at a resistance of 2 Ω between the electrodes, was obtained during an experimental period of 7 days.

Essentially the same procedure is described in the latest article by Karube et al.[16] Improvements include that the agar immobilized organisms are carried on acetyl cellulose filters which increased the hydrogen production rate due to less diffusion resistance in the matrix. Continuous production by means of repeated batches was performed in a stirred fermenter containing 2 kg of immobilized cells (a total of 400 g cells wet weight) and 3 *ℓ* of waste water (BOD 10000 ppm). Feeding with condensed waste water was necessary to ensure good hydrogen production (10 m*ℓ*/min, BOD 70000 ppm). After an initial period, during which the activity increased due to growth of organisms in the gel, 54 mmol H_2/h·kg gel (or 0.27 mmol/h·g wet cells originally immobilized) was obtained for 30 days. Hydrogen gas, at a flow rate of 40 m*ℓ*/min, was then transferred to five fuel cells connected in a series. A steady-state current of 0.8 A, at a total resistance of 2 Ω between the electrodes, was obtained in a 7-day long experiment (Figure 4).

There seems to be a number of different problems involved in the production of electrical energy from waste water. There are all the biological problems involved in the production of hydrogen but also the problems in the effective conversion of hydrogen to electrical energy. In the biological system the ultimate goal must be to reach a high degree of conversion of substrate to product with a high reaction rate. Important parameters are then the organisms specific rate of hydrogen production, the biomass concentration in the reactor, and the yield coefficient for hydrogen. For immobilized cells the specific rate of product formation that can be measured is only an apparent value. It is influenced by a number of factors, not only biological ones. Among non-biological factors the immobilization method, the matrix used, and the reactor design influence survival of the organism, mass transport of substrate and products, mechanical strength of the aggregates, and cell leakage. All these factors result in an unknown amount of active biomass exposed to a varying environment. If the cells metabolic capacity is to be utilized fully the external environment, regarding nutrients, pH, temperature, and in this case also the presence or absence of oxygen, is of course very important.

The papers cited above do not give enough data to enable a detailed investigation of how different experimental conditions influence the final results. It is possible though to conclude that the system with continuous medium flow through a packed bed reactor gave the best results regarding both specific rate of hydrogen production (0.6 mmol/h·g wet cells originally immobilized) and total productivity (26.6 mmol/h·ℓ reactor-vol.).[14] When continuous conditions were simulated by repeated batches in a stirred fermenter the specific rate of hydrogen production decreased (0.14 to 0.27 mmol/h·g wet cells) as did also the productivity (2.9 to 21.6 mmol/h·ℓ reactor vol).[15,16] One explanation to these differences is certainly that the fragile agar gel is not mechanically stable enough to stand the stirring but there might be others as well.

Then comes the problem with the effective conversion of hydrogen gas to electrical energy in the fuel cell. This is obviously not a microbiological problem and will therefore be discussed only briefly here. It seems reasonable to assume that the development of a fuel cell preferentially should be performed separately from the biological system. An example can be supplied by comparing the results from the last two articles cited above. In the first case two fuel cells were connected in a series and supplied with pure hydrogen at a flow rate of 6 mℓ/min.[15] Knowing the current produced (500 to 550 mA) and the resistance between the electrodes (2 Ω) the total effect can be calculated to 0.605 W. In the last article five cells, apparently of the same construction as earlier, were connected in a series and supplied with hydrogen at 40 mℓ/min.[16] Similarly the effect can be calculated to 1.28 W. Thus, the hydrogen supply was increased 6.7 times but the effect only doubled. If the amount of biomass used to create the energy also is taken into account it can be calculated that in the first case 24.2 W are produced per kg dry weight of cells and 12.8 W per kg dry weight of cells in the second case. The concept of bio fuel cells is interesting but obviously a substantial improvement of the fuel cell is required.

Finally, some general remarks on the practical application of this system. When the waste water has been utilized for hydrogen production it still contains considerable amounts of organic substances, although some carbon dioxide has been released. Further BOD reduction, either anaerobically or aerobically is required before it can be disposed of. Thus, this process does not contribute substantially to the cleaning of the waste water. By anaerobic digestion both BOD reduction and energy in the form of methane gas can be obtained in the same process. However, electrical energy is attractive in that it is a very clean source and it would probably not be wise to rule out this possibility before the potential of the system has been evaluated fully. Another very important question in the application of this system to full scale is the maintenance of a culture with a specific metabolic activity in a process that has to be operated at nonsterile conditions.

IV. ACETONE, BUTANOL, AND ISOPROPANOL

The acetone-butanol fermentation is one of the oldest microbial processes on an industrial scale and was also one of the most important ones. Before World War II there was no other method available for the production of the desired products. With the development of the petrochemical industry most microbial production plants were closed. Today, this old process is again attracting interest because of the possibility to produce solvents from renewable resources. Economic calculations reveal though that it is not yet feasible to produce acetone and butanol by fermentation from substrates like molasses and starch.[17]

The main limitations of the microbial process are the low productivity (7 to 9 g butanol/ℓ·day), the low maximum solvent concentration (2%), the low yield of products (30 to 35%).[18,19] Some of these problems can be solved by a more efficient process technology. The old industrial process was run batchwise with much time and labor required for starting up and cleaning of the equipment. Attempts were made to run the acetone-butanol process continuously in pilot plant scale, when techniques for continuous culture were developed in the fifties.[20] The organisms used were *Clostridium acetobutylicum* for production from starches, or saccharolytic clostridia, mainly described in the patent literature under a large number of different names, for production from sugars.[21,22] In a batch culture, two distinct phases can be distinguished.[23] In the first phase the organisms grow logarithmically, producing acetic and butyric acids which also cause a decrease in the culture pH. In the second phase growth ceases and the already produced acids are converted into solvents but acetone, butanol, and ethanol are also produced directly from the carbon source. Our knowledge about the metabolic events is still incomplete and much work remains to be done especially regarding the biochemical regulatory mechanisms.

However, by the application of nongrowing immobilized cells to the production of acetone and butanol several obvious advantages would be achieved: (1) if the same cells are used for longer time periods a higher product yield should be obtained since less of the substrate is consumed for biomass production; (2) continuous operation at high cell densities with dilution rates exceeding the growth rate would be possible, thereby leading to higher productivity; (3) the products can be separated from the medium without damage to the cells and the substrate recycled in the process; (4) the substrate used can be simplified since it does not need to contain nutrients required for growth; (5) waste biomass will generate less problems since most cells are retained in the reactor; (6) the system will be less sensitive to oxygen since the organism can maintain its own reduced microenvironment.

The feasibility of such a system was investigated by Häggström, and Häggström and Molin.[24,25] Initially it was necessary to establish how *Clostridium acetobutylicum* behaved under nongrowth conditions. Organisms from the logarithmic growth phase were immobilized in a calcium alginate gel and incubated in a nongrowth substrate containing glucose and inorganic salts. The product formation pattern was similar to that of normal batch grown cells as the two biological production phases could be distinguished. The results were also verified using resting suspended cells.

Economical use of immobilized cells require continuous operation or repeated use in a batch process. It was assumed that for continuous acetone-butanol production a two-stage process would be necessary with acid production in the first step followed by solvent production from the acids and the carbohydrate source in the second step. Consequently, the metabolic behavior of phase-two cells were investigated. A suspension of washed resting cells were incubated in media of various composition. These cells produced acetone and butanol directly from a glucose-salts medium. If the substrate contained glucose and butyric acid the butanol production rate was doubled due to the direct conversion of butyric acid to butanol. With acetic acid, butyric acid, and glucose in the medium the result was virtually the same as with glucose only. It thus seemed possible that by using immobilized cells an one-step process could be designed either with glucose as the sole substrate or with a mixture of glucose and butyric acid.

During the course of the experimental work it became clear that vegetative phase-two cells were extremely sensitive to not only oxygen, but also factors like temperature changes and glucose starvation. In other words, it was rather difficult to obtain an active preparation of immobilized cells. To overcome these problems spores instead of vegetative cells were immobilized in the calcium alginate gel. By using spores oxygen was no longer any problem, nor was it necessary to minimize the time required for the hardening of the gel. Immobilized spores can be stored for long times. Spores are also more resistant to chemicals than are vegetative cells. In order to increase the gel strength, glutardialdehyde was added to the calcium chloride solution containing the newly formed gel beads. This mixture was stored overnight without any loss of viability in the spores. Before further use, the gel beads were washed carefully. The concept of immobilizing spores is very attractive when sensitive anaerobes are used since many practical problems in keeping the system anaerobic are avoided. A wider range of immobilization methods are also available, which otherwise could have been detrimental to the vegetative cells.

The spores in the gel were then heat shocked by the addition of 95°C hot growth medium and incubated until the vegetative cells, formed inside the gel matrix, had entered the solvent production phase. For continuous production the washed gel beads (Ø 2mm), prepared from 100 g alginate-spore solution, were transferred to a column (125 mℓ volume) that was connected to a laboratory fermenter in such a way that the substrate was pumped at a high rate from the fermenter through the column and back to the fermenter again (Figure 5).[26] The purpose of this procedure was to obtain a homogeneous distribution of substrate and products in the column. Fresh nongrowth medium (137 mℓ/hr) was added directly to the fermenter where the liquid level (2500 mℓ) also was controlled. The spent substrate leaving the reactor contained 2.05 g butanol/ℓ, 0.37 g acetone/ℓ and 0.7 g acetic acid/ℓ during these conditions. Simultaneously the organisms had consumed 6.75 g glucose/ℓ, and 1.11 g butyric acid/ℓ. The butanol production rate was calculated to 67 g butanol/1000 g alginate solution·day. The alginate gel occupies approximately the same volume as the original alginate solution which means that in a packed bed reactor the productivity would be 67 g butanol/ℓ reactor-volume·day. The specific rate of butanol production has not been calculated because the concentration of bacterial dry weight in the gel was not known and also very difficult to estimate. However, in a long-time experiment it turned out that a rather rapid activity loss occurred after the initial period, but after that production could go on for long times (1000 hr) at a relatively low (10 to 20 g butanol/ℓ·day) but steady level (Figure 6). The reason for this complex behavior is not quite clear but the results indicate that long time use of this system is fully possible.

In order to improve the experimental results, the design of the reactor and the problem with maintenance of activity was considered. Columns packed with spherical gel beads are actually not suitable for immobilized cell systems where large amounts of gas are produced. Even in small units problems with gas pockets and channeling are encountered. It also seems very difficult to scale up such a system. Attractive alternatives are then fluidized beds or stirred fermenters but gels like polyacrylamide, agar, and alginate are seldom stable enough to stand the mechanical stress in the long run. In the author's laboratory we have recently tried a new approach to solve these problems. The reactor consists of a cylindrical glass vessel containing concentric stainless steel netcylinders. The spore containing gel is immobilized as a thin film on the wire netting and the substrate is mixed by a stirrer to assure homogeneity in all parts of the reactor (Figure 7a). The gas leaves the reactor without problems and all parts of the gel surface is exposed to the same environment.

Nongrowing cells will inevitably lose activity by time if nutrients are not supplied. This is also valid for immobilized cells of *Clostridium acetobutylicum*. A technique for maintenance of constant activity of immobilized nongrowing cells has therefore been developed. Doses of nutrients are distributed to the reactor at intervals. The nutrient pulse should be sufficient to enable the organism to build up essential cell constituents but not enough to allow multiplication to proceed. Figure 7b shows the ex-

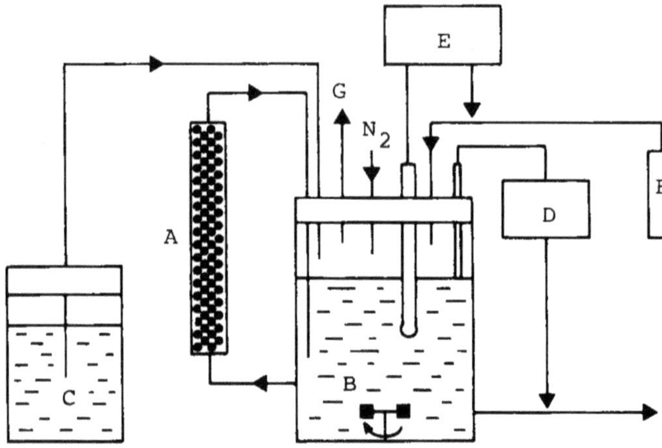

FIGURE 5. System for continuous production of acetone and butanol by alginate immobilized nongrowing cells of *Clostridium acetobutylicum*. (A) Column with immobilized cells. (B) Fermenter. (C) Reservoir for nongrowth solution. (D) Liquid level control. (E) pH-control. (F) Sodium hydroxide reservoir. (G) Gas outlet.

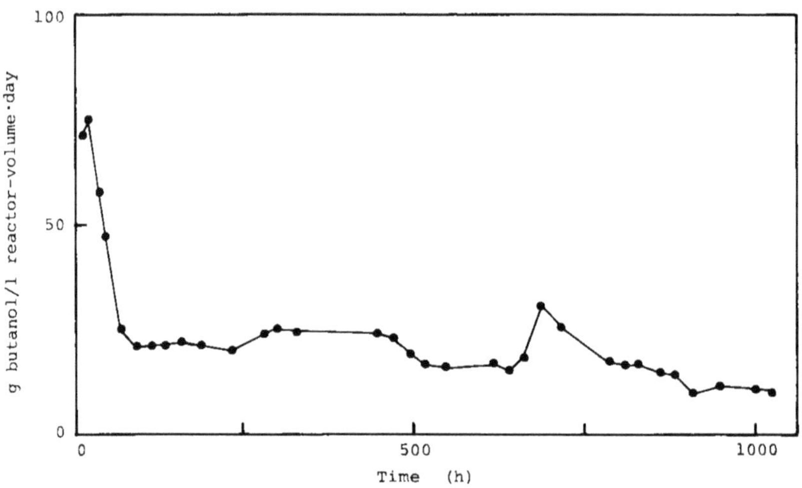

FIGURE 6. Butanol production by alginate immobilized *Clostridium acetobutylicum*.

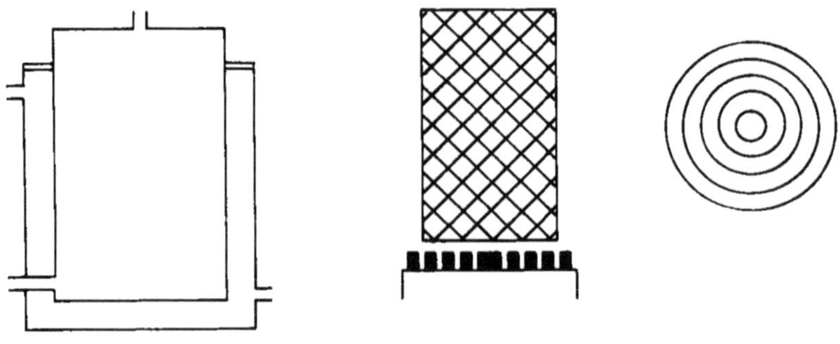

FIGURE 7a. Reactor for immobilized cells. (A) Glass vessel. (B) Cylindrical wire netting. (C) Cross-section of the cylinders.

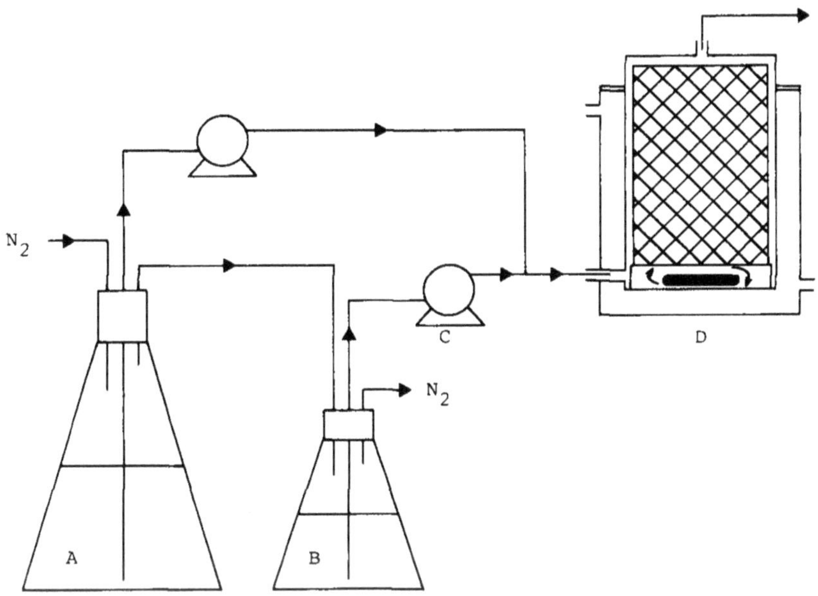

FIGURE 7b. Experimental set-up for continuous production of acetone and butanol with the nutrient dose technique. (A) Nongrowth solution. (B) Nutrient substrate. (C) Pump for pulse wise addition of nutrients. (D) Reactor with immobilized cells.

perimental set-up. In experiments performed in the reactor described above, using the nutrient dose technique, an initial activity loss was, however, repeatedly encountered. Further investigations showed this to be caused by the diffusion resistance in the gel matrix. The system could not be controlled until a surface layer of organisms were formed by rich nutrient supply.

Figure 8 shows the result from an experiment where nutrient pulses initially were distributed at short intervals to force the organisms to grow and form a surface layer. The dose interval was then increased slowly. As judged by the absence of free cells in the medium growth was restricted at a dose interval of 8 hr. At the same time a constant productivity of butanol was maintained. During these conditions the yield of butanol has been calculated to 20% (w/w) of the glucose consumed.

Based on present knowledge of the metabolism in *Clostridium acetobutylicum* the maximum theoretical yield of butanol is estimated to 27% (w/w). The amount of butanol produced is restricted by the supply of reducing power in the cell. For production of 1 mole of butanol 4 moles of NADH are required. In the break down of glucose 2 moles NADH are produced per mole of glucose in the glycolysis. Varying amounts of NADH and hydrogen gas are also formed in the phosphoroclastic split of pyruvate. Based on published data on the hydrogen production the NADH produced can be estimated to 0.65 mol/mol glucose.[9] Thus, a total of 2.65 mol NADH/mol glucose can maximally give 0.66 mol butanol/mol glucose. Only if conditions for a higher production of NADH can be found this figure will be increased.

Thus, these results show that nongrowing immobilized cells provide conditions for obtaining reasonably high possible yields of products. In the experiment shown in Figure 8, totally 25 g of alginate solution was immobilized on the wire netting (total area 690 cm^2) in a reactor with 175 mℓ liquid volume. At a dilution of 0.4 hr^{-1} the product stream contained 1.75 g butanol/ℓ. The volumetric productivity was thus lower (16.8 g butanol/$\ell \cdot$day) compared to the packed bed reactor because the amount of biomass per volume unit was decreased due to a larger void volume. The butanol production rate per g alginate solution was, however, higher (140 g butanol/1000 g alginate

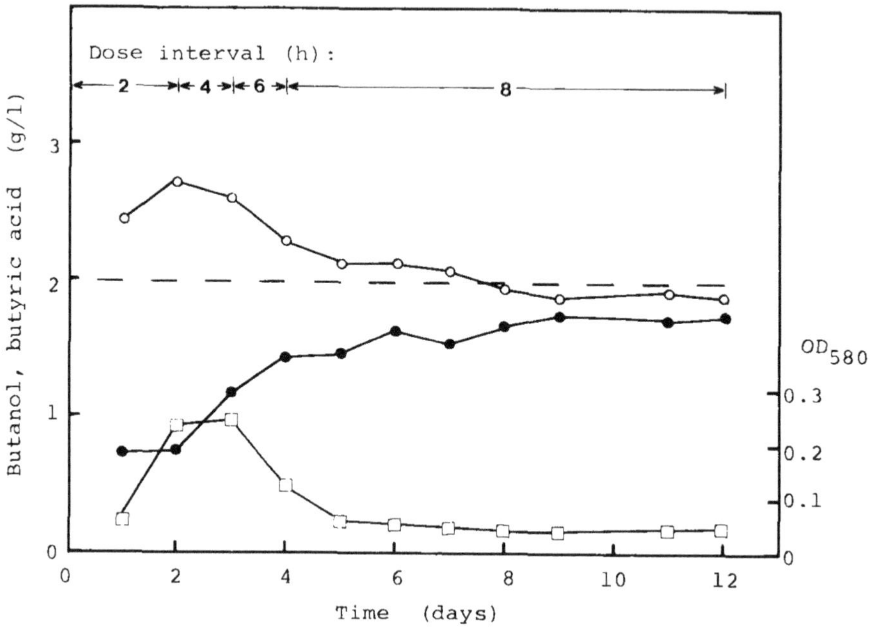

FIGURE 8. Continuous production of butanol by immobilized cells of *Clostridium acetobutylicum* using the nutrient dose technique. Butanol (—●—), butyric acid (—○—), and OD_{580} (—□—). The dotted line indicates the medium level of butyric acid.

solution·day) indicating that the gel was utilized more effectively in this construction. However, it is rather the surface area that is important and not the amount of gel. Developments will therefore be focused on more technically and economically interesting support materials for the adsorption of microorganisms.

Krouwel et al.[27] have investigated the production of butanol and isopropanol from the related organism *Clostridium butylicum* with a similar system. Spores and cell debris from a 48-hr culture were immobilized in a calcium alginate gel. The spherical particles (1 mm) were kept in a conical column with approximately 30 mℓ liquid volume. A continuous medium flow of 8.6 mℓ/hour ($D = 29\ hr^{-1}$) was applied. The medium contained glucose and yeast extract and was thus a complete growth medium. During these conditions the organism will grow and free cells are liberated into the medium. After an initial adaptation period 5 g butanol/ℓ and 3 g isopropanol/ℓ were formed at a steady-state level, the total yield of solvents being approximately 30% and the productivity 24 g butanol/ℓ·day according to the authors. Approximately 1 g dW of cells/ℓ was found in the outflowing medium. However, after 130 hr of continuous production the concentration of solvents began to drop. The authors believe that this might be an effect of the low pH developed during the process.

V. STEREOSPECIFIC HYDROGENATIONS

One application for immobilized cells is to use them in a synthesis chain to bring about a chemical transformation of the substrate which otherwise is very complicated or impossible to do by chemical synthesis. An example on such transformations is stereospecific hydrogenations performed by clostridia in the presence of hydrogen and a suitable hydrogen acceptor. Tischer et al.[28] have studied the stability of free and immobilized cells of *Clostridium* DSM 1460 used for the reduction of (E)-2-methyl-2-butenoate. The enzymes involved in the reduction would be: hydrogenase, NADH-ferredoxin-oxidoreductase, and enoate reductase. Immobilization of the cells in polyacrylamide greatly reduced the microbial activity. Of the ingredients used acrylamide and ammonium persulfate were the most damaging agents while N,N,N',N'-tetramethyl-ethylene-diamine had no adverse effect and N,N'-methylene-bis-acrylamide reduced the activity only slightly. Addition of 0.8-1% dithionite to the reaction mixture reduced the negative effects of the other components. In a hydrogenation experiment the initial reduction rate of the enoate, as measured by the H_2-uptake, was highest for the free cells (30 μmol $H_2 \cdot 0.1$ g wet packed cells (WPC) intermediary for gelatin-formaldehyde immobilized cells (14 μmol $H_2/h \cdot 0.1$ g WPC) and lowest for polyacrylamide immobilized cells (7 μmol $H_2/h \cdot 0.1$ g WPC). The half-life for both free and immobilized cells was in the range of 100 to 200 hr. In spite of the better activity for the free cells there was one great advantage with the immobilized cells, namely that they could be reused repeatedly. When free cells were separated from the reaction mixture the sediment was completely inactive but a reduced activity was found in the supernatant indicating that the cells were lysed and the enzymes liberated into the solution. These results indicate the possible use of immobilized anaerobic organisms for special purposes. The system is rather attractive in that it operates under very clean conditions: a buffer solution, hydrogen, and the substrate to be transformed. It certainly requires further development but the conditions seem promising as the gelatin entrapped organisms maintained a rather steady activity for 600 hr after an initial loss of activity.

VI. CONCLUSIONS

Most of the systems described here have the common property of being anaerobic energy conversion processes. Which demands must such systems fulfill and to what extent does the immobilization in itself play an important role?

Anaerobic metabolism yields very little biomass in relation to the amount of substrate consumed by the organism. Anaerobic processes are therefore generally characterized by a low productivity. If these processes are to be used for the production of fuels or bulk chemicals it is essential to increase the biomass concentration per volume unit of the reactor. There are several methods available to achieve this goal, e.g., sludge formation as it is used in waste water treatment, recirculation of biomass by means of a separation step, and the use of immobilized cells. If microorganisms are entrapped in a polymeric network or adsorbed to solid surfaces under growth conditions, newly formed cells will be concentrated and retained in the reactor. Besides this advantage the immobilization makes the system more stable and less sensitive to disturbances which otherwise could have led to a wash-out situation. In the immobilized state anaerobic microorganisms can maintain their own anaerobic microenvironment and the handling of these sensitive organisms are facilitated.

Other important factors in energy conversion processes are the yield coefficient for the product and the product concentration. This might not be so essential in the case of methane or hydrogen production if we assume that these products are going to be produced from waste waters. For products like acetone, butanol, and isopropanol it is though of utmost importance. The raw materials for these products are sugars, starches, and cellulosic hydrolysates which have a high price in itself. A high yield coefficient is necessary for the microbial process to be economic. Here, the ultimate limitation lies in the biological system and the immobilization does not help much unless nongrowing cells are applied. In that way the metabolic capacity of the cells is utilized fully and part of the substrate which otherwise would have been spent for biomass production is saved for product formation.

In what way can immobilized processes help to bring up the product concentration? It is unlikely that the immobilization in itself will influence the product concentration but it may provide conditions for more effective separation methods. Extractive fermentation has been used in different processes. The main problem is that most common extractants also are inhibitory for the microorganisms. By choosing the immobilization matrix carefully, it might not be impossible to create a system where the extractant has limited access to the microorganism and its microenvironment. Another idea can be the use of permeabilized cells which possibly can result in a higher extracellular product concentration since high intracellular product concentrations are suspected to limit the product tolerance.[29] In doing this, the use of immobilized cells would be an advantage since permeabilization means expensive treatment of the cells.

Generally, in microbial processes for the production of cheap bulk products the economy is very strained. The systems described here fall into this category. It is rather unlikely that one can afford to use such sophisticated and expensive carriers as agar, polyacrylamide, and alginate on a large scale. These methods are neither suitable for scaling up. It must also be taken into account that the carrier to be used must be available in large quantities, be reasonably cheap, and can be repeatedly used. Research in this area should therefore be oriented to meet these demands. Thus, much work remains to be done before it is possible to make a fair economic evaluation of the prospects of immobilized cell processes for the production of fuels and chemical feedstocks.

REFERENCES

1. **Gottschalk, G.**, *Bacterial Metabolism*, 2nd ed., Springer-Verlag, New York, 1979, 167.
2. **Mattiasson, B., Ed.**, *Immobilized Cells and Organelles*, CRC Press, Boca Raton, Fla., 1983.
3. **Genung, R. K., Million, D. L., Hancher, C. W., and Pitt, W. W., Jr.**, Pilot plant demonstration of an anaerobic fixed-film bioreactor for wastewater treatment, in *Biotechnology Bioengeering Symposium No. 8*, Scott, C. D., Ed., John Wiley & Sons, New York, 1978, 329.
4. **Frostell, B.**, Wastewater treatment and energy recovery in an expanded bed system, in Proc. 3rd Int. Congr. Industrial Waste Water and Wastes, Stockholm, Feb. 6 to 8, 1980.
5. **Karube, I., Kuriyama, S., Matsunaga, T., and Suzuki, S.**, Methane production from wastewaters by immobilized methanogenic bacteria, *Biotechnol. Bioeng.*, 22, 847, 1980.
6. **Mattiasson, B.**, Biochemical applications and perspectives of immobilized multistep enzyme systems, in *Biomedical Applications of Immobilized Enzymes and Proteins*, Vol. 2, Chang, T. M. S., Ed., Plenum Press, New York, 1977, 253.
7. **Hansson, G.**, Effects of carbon dioxide and methane on methanogenesis, *Eur. J. Appl. Microbiol. Biotechnol.*, 6, 351, 1979.
8. **Scherer, P., Kluge, M., Klein, J., and Sahm, H.**, Immobilization of the methanogenic bacterium *Methanosarcina barkeri*, *Biotechnol. Bioeng.*, 23, 1057, 1981.
9. **Wood, W. A.**, Fermentation of carbohydrates, in *The Bacteria*, Vol. 2, Gunsalus, I. C. and Stanier, R. Y., Eds., Academic Press, New York, 1961, chap. 2.
10. **Karube, I., Matsunaga, T., Tsuru, S., and Suzuki, S.**, Biochemical fuel cell utilizing immobilized cells of *Clostridium butyricum*, *Biotechnol. Bioeng.*, 19, 1727, 1977.
11. **Karube, I., Matsunaga, T., Tsuru, S., and Suzuki, S.**, Continuous hydrogen production by immobilized whole cells of *Clostribium butyricum*, *Biochem. Biophys. Acta*, 444, 338, 1976.
12. **Matsunaga, T., Karube, I., and Suzuki, S.**, Some observations on immobilized hydrogen-producing bacteria: behaviour of hydrogen in gel membranes, *Biotechnol. Bioeng.*, 22, 2607, 1980.
13. **Karube, I., Matsunaga, T., Mitsuda, S., and Suzuki, S.**, Microbial electrode as BOD sensors, *Biotechnol. Bioeng.*, 19, 1535, 1977.
14. **Suzuki, S., Karube, I., and Matsunaga, T.**, Application of a biochemical fuel cell to wastewaters, in *Biotechnology Bioengeering Symposium No 8*, Scott, C. D., Ed., John Wiley & Sons, New York, 1978, 501.
15. **Suzuki, S., Karube, I., Matsunaga, T., Kuriyama, S., Suzuki, N., Shirogami, T., and Takamura, T.**, Biochemical energy conversion using immobilized whole cells of *Clostridium butyricum*, *Biochimie*, 62, 353, 1980.

16. **Karube, I., Suzuki, S., Matsunaga, T., and Kuriyama, S.**, Biochemical energy conversion by immobilized whole cells, *Ann. N.Y. Acad. Sci.*, 369, 91, 1981.
17. **Bronn, W. K.**, Verfahrenstechnische Probleme und Wirtschaftlichkeit der garungstechnischen Produktion von Butanol-Aceton, in *Proc. 4. Symp. Technische Mikrobiologie*, Dellweg, H., Ed., Verlag Versuchs- und Lehranstalt fur Spriitusfabrikation und Fermentationstechnologie im Institut fur Garungsgewerbe und Biotechnologie, Berlin, 1979, 381.
18. **Prescott, S. C. and Bunn, C. G.**, *Industrial Microbiology*, 3rd ed., McGraw-Hill, New York, 1959, chap. 13.
19. **Spivey, M. J.**, The acetone/butanol/ethanol fermentation, *Proc. Biochem.*, 13(11), 2, 1978.
20. **Hospodka, J.**, Industrial application of continuous fermentation, in *Theoretical and Methodological Basis of Continuous Culture of Microorganisms*, Malek, I. and Fencl, Z., Eds., Publishing House of the Czechoslovak Academy of Sciences, Prague, 1966, 611.
21. **Beesch, S. C.**, Acetone-butanol fermentation of sugars, *Ind. Eng. Chem.*, 44, 1677, 1952.
22. **Beesch, S. C.**, Acetone-butanol fermentation of starches, *Appl. Microbiol.*, 1, 85, 1953.
23. **Peterson, W. H. and Fred, E. B.**, Butyl-acetone fermentation of corn meal. Interrelations of substrate and products, *Ind. Eng. Chem.*, 24, 237, 1932.
24. **Häggström, L.**, Production of acetone and butanol with immobilized cells of *Clostridium acetobutylicum*, in *Proc. 4. Symp. Technische Mikrobiologie*, Dellweg, H., Ed., Verlag Versuchs- und Lehranstalt fur Spriitus-fabrikation und Fermentationstechnologie im Institut fur Garungsgewerbe und Biotechnologie, Berlin, 1979, 271.
25. **Häggström, L.**, Calcium alginate immobilized cells of *Clostridium acetobutylicum* for solvent production, *Biotechnol. Lett.*, 2, 241, 1980.
26. **Häggström, L.**, Immobilized cells of *Clostridium acetobutylicum* for butanol production, in *Advances in Biotechnology*, Vol. 2, Moo-Young, M. and Robinson, C. W., Eds., Pergamon Press, Toronto, 1981, 79.
27. **Krouwel, P. G., van der Laan, W. F. M., and Kossen, N. W. F.**, Continuous production of n-butanol and isopropanol by immobilized, growing *Clostridium butylicum* cells, *Biotechnol. Lett.*, 2, 253, 1980.
28. **Tischer, W., Tiemeyer, W., and Simon, H.**, Stereospecific hydrogenations with immobilized microbial cells or enzymes, *Biochimie*, 62, 331, 1980.
29. **Novak, M., Strehaiano, P., Moreno, M., and Goma, G.**, Alcoholic fermentation: On the inhibitory effect of ethanol, *Biotechnol. Bioeng.*, 23, 207, 1981.

Chapter 5

CO-IMMOBILIZATION INVOLVING CELLS, ORGANELLES, AND ENZYMES

Bärbel Hahn-Hägerdal

TABLE OF CONTENTS

I. INTRODUCTION

Microorganisms often appear immobilized in nature in the sense that they naturally adhere to solid supports in their environment. In most instances these naturally occurring immobilized microorganisms do not exist as a monoculture, but rather as an ecologically well balanced mixed culture. In other words, many microorganisms exist co-immobilized in nature.

Technically this phenomenon has been exploited for a long time in biological waste water treatment, where undefined mixed cultures which adhere to various kinds of support, are used to reduce the amount of organic material in the waste water stream.

However, co-immobilization involving defined biocatalytic species is a relatively unexplored area. Co-immobilization involving microbial cells may be regarded as an extension of the use of mixed cultures. The current status of the use of mixed cultures in biotechnical processes has recently been reviewed[1] and it appears that the only area in which mixed cultures so far have had real technical success is in the field of single cell production. Other areas in which the author envisions that mixed cultures may have a potential is in the production of vitamin B_{12}, organic acids, metal leaching, and steroid transformation.

The interactions between microbial species have been characterized into seven categories:[2] competition, predation, parasitism, amensalism, neutralism, commensalism, and mutualism. Stable mixed cultures are most likely to occur only with the last three types of interactions. Interactions between other biocatalysts such as enzymes, organelles, and cells might be described in a similar way. Thus translating microbial species with biocatalysts the stable interactions might be described as:

- *Neutralism:* Two biocatalytic species function in the same environment without interaction.
- *Commensalism:* The production of a growth factor (substrate) by one biocatalyst which is essential and can be consumed by the second biocatalyst.
- *Mutualism:* Both biocatalytic species produce growth factors (substrates), which are essential and can be consumed by the other.

Immobilization of biocatalysts offers several advantages over the utilization of free biocatalysts:

- High catalytic density with concomitant increased productivity is achieved.
- The catalysts can easily be recovered.
- The immobilized system can be operated continuously.
- Upon immobilization biocatalysts often gain stability against thermal as well as environmental denaturation.
- Due to the fact that immobilization creates a phase separation between product and catalyst already in the conversion process, cleaner product streams are obtained, which facilitates product upgrading.
- Co-immobilization of two or more biocatalytic species brings the catalysts in close proximity to each other, which results in higher productivity of the overall system.[3]
- In a continuous mixed culture, wash out at high dilution rates might be avoided and the system could be operated with increased productivity.

In the following text, co-immobilization will be discussed in relation to the kind of biocatalytic species involved starting with mixed undefined cultures. The following topics will be subsequently discussed: the co-immobilization of two organisms; one organism and an organelle; one organism and an enzyme, and finally an organelle and an enzyme.

II. IMMOBILIZED UNDEFINED MIXED CULTURES

Both aerobic and anaerobic waste treatment utilize attached mixed cultures. For aerobic treatment trickling filters and rotating disks have been used in order to improve aeration. About 50% of the organic carbon of the waste is converted into cell mass in an aerobic treatment, whereas 5% is converted to cell mass and another approximately 80% to a mixture of carbon dioxide and methane in an anaerobic treatment. The latter treatment of waste seems to have more bearing for the future since it is a more efficient process. In spite of this fact, anaerobic treatment of domestic waste water, which is considered low-strength waste in that the biological oxygen demand (BOD) is less than 500 mg/ℓ, is not regarded as an efficient process, the main reason being that it is difficult to retain a sufficient quantity of cell mass in the reactor. One way to overcome this problem is to use an "anaerobic filter" process, where the waste water stream is led upwards through a bed of inert support to which sludge is attached.[4-6] Various support materials have been studied: aluminum oxide, crude diatomaceous silica, calcium carbonate, sand, and polyurethane. The area to volume ratio, the toxicity, and the price of the support material are important features that have to be considered when choosing support material for an "anaerobic filter" waste treatment process.

Anaerobic treatment of waste is based on a commensal type of interaction between three metabolic groups of bacteria.[7,8] The first group excrete enzymes which hydrolyze lipids, proteins, and carbohydrates. The hydrolysis products are fermented to fatty acids, carbon dioxide, and hydrogen which are subsequently excreted. The second group, often called acetogenic bacteria, produce hydrogen and acetate from the endproducts produced by the first group of bacteria. Finally, the methanogenic bacteria produce methane and carbon dioxide from the endproducts produced by the first two groups of bacteria.

The microbial culture is immobilized in the support material by seeding it, after which the microorganisms grow over the available surface area. Thus, the natural ability of the microbial biomass that encompasses the anaerobic sludge to adhere to and grow on solid supports is used as immobilization technique.

In laboratory experiments the "anaerobic filter" has been very successful for treating low-strength waste.[5] The success is mainly ascribed to the large surface to volume ratio created by the inert support material which enables a large active mass of attached microorganisms to remain in the reactor at high liquid flow rates. In other words the advantage of the immobilization in this case is due to the high catalytic density obtained in the reactor.

So far the "anaerobic filter", despite successful laboratory reports, has not been exploited in a technical scale. The main reason seems to be the price for a good nontoxic support material.

Anaerobic sludge inherently has superior settling characteristics, provided that the physical and chemical conditions for sludge flocculation are favorable. This feature is the basis for the "upflow anaerobic sludge blanket" process, whereby a high biomass concentration is maintained by recycling the flocculated sludge.[9] Flocculation might also be regarded as a kind of immobilization, in the most general sense of the word. Similar to anaerobic waste treatment, the efficiency of denitrification can be highly improved by using attached sludge whereby a high biocatalytic density is obtained.[10]

The complete removal of nitrogen from water by microbes, a process involving both nitrification and denitrification, is also due to cooperative interactions between several bacterial species: *Nitrosomonas,* which oxidize ammonia to nitrite and *Nitrobacter* which further oxidize the nitrite to nitrate. From here, nitrate might again be reduced to molecular nitrogen by several *Pseudomonas sp.*

Another approach to the anaerobic waste water treatment has been to entrap a methanogenic microbial population from digested industrial sewage sludge in various carrier systems.[11] Agar, polyacrylamide, and collagen have been investigated. Microorganisms immobilized in agar gave the highest retained activity, which, however, was only half of the initial activity of the free cells. Nevertheless, the immobilized microorganisms had several advantages over the free microbial culture. First of all, the immobilized cells could continuously produce methane under aerobic conditions, due to the fact that the carrier gel created a diffusion barrier which prevented the oxygen molecules from reaching the anaerobic bacteria. It was also found that the total amount of methane produced by the immobilized microorganisms was larger than that of the free cells. This was attributed to the fact that the products formed by the hydrolytic and acetogenic bacteria could be more efficiently utilized by the methanogenic bacteria since diffusion restrictions in the carrier gel caused the products to be enriched in the vicinity of these microorganisms. Even though immobilization of a methanogenic microbial population in a carrier gel provides these advantages, it is questionable whether or not a waste water treatment process can economically carry the cost of such an immobilization. The system might rather be useful for fundamental studies on microenvironmental effects on an immobilized mixed culture, which might be more difficult to study on an attached sludge system.

Co-immobilized cells have also been used for rapid BOD estimations in waste water streams. An undefined mixed bacterial culture isolated from soil was immobilized in polyacrylamide. The bacteria were subsequently exposed to a sample, which was saturated with oxygen (Figure 1).[12] An oxygen electrode, which was immersed in this sample, showed that the oxygen consumption was proportional to the BOD for values up to 350 ppm. This method of BOD determination has several advantages over other methods in that the determination time is very short (30 min) and that the system is reusable. Over a 10-day period, no decrease in the respiratory activity of the immobilized bacteria was reported. The main disadvantage of this method, however, is that it can only be used for waste waters that contain organic compounds which the soil bacteria can degrade. The authors suggest that microorganisms isolated from activated sludge should be used for BOD estimations of waste waters since these bacteria can degrade the organic compounds which the soil bacteria cannot.

A curiosity among the undefined immobilized mixed cultures is one which had been derived from kefir. It contained lactobacilli and lactose fermenting yeast for the production of lactic acid from waste whey.[13] The mixed culture was immobilized by seeding a support material described as "berl saddles". The main rationale for using a mixed culture in this application was that a culture derived from kefir would be stable against phage infection and to storage and reconstitution. This stability was thought more important than the higher productivity of lactic acid, which could be achieved with a homo fermentative lactobacillus strain. None of these considerations was proven experimentally, however. The reason for using an immobilized culture was to design a continuously operating system for waste whey treatment.

III. CO-IMMOBILIZED MICROORGANISMS

Martin and Perlman[14] co-immobilized two microorganisms, *Gluconobacter melanogenes* IFO 3293 and *Pseudomonas syringae* NRRL B-865, in polyacrylamide for the conversion of L-sorbose to 2-keto-L-gulonic acid.

G. melanogenes alone can convert L-sorbose to 2-keto-L-gulonic acid with L-sorbosone as an intermediate. This conversion appears to be limited in the second enzymatic step catalyzed by L-sorbosone oxidase. It was also found that L-sorbosone produced by separately immobilized *G. melanogenes* inactivated the L-sorbose dehydrogenase, which might be a result of diffusion limitations within the polyacrylamide gel used for the immobilization.

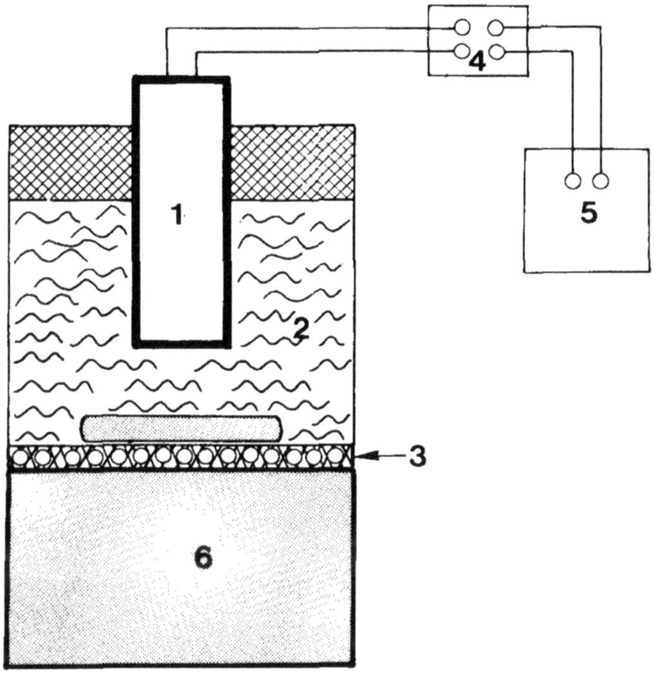

FIGURE 1. Schematic diagram of a system to estimate BOD. 1. Oxygen electrode, 2. Sample solution, 3. Immobilized microorganisms, 4. Amplifier, 5. Recorder, 6. Magnetic stirrer. (From Karube, I., et al., *J. Ferment. Technol.*, 55(3), 243, 1977. With permission.)

In order to overcome these problems, a strain of *P. syringae* having a high L-sorbosone oxidase activity was co-immobilized with *G. melanogenes*, resulting in an immobilized commensual-type mixed culture. When the co-immobilized preparation was compared with the preparations of separately immobilized bacteria, it produced 3,100 μg/mℓ of 2-keto-L-gulonic acid, whereas *G. melanogenes* itself only produced 730 μg/mℓ and *P. syringe* produced none from L-sorbose.

The authors also report that the L-sorbose dehydrogenase activity seems to be stabilized in the co-immobilized preparation probably due to the fact that the product is rapidly converted to 2-keto-L-gulonic acid. No comparison was made between the co-immobilized preparation and a mixture of the separately immobilized bacteria in order to elucidate the effects of the proximity achieved by co-immobilizing. The major drawback of this system was the instability of the L-sorbosone oxidase at 45°C, which was the optimum temperature for the conversion of L-sorbose to L-sorbosone.

This illustrates the greatest and most fundamental problem of co-immobilizing two biocatalytic species which have not been ecologically adapted to the same environment. It also stresses the major limitation of co-immobilizing two biocatalysts: the chain is not stronger than the weakest link. In order to solve this problem, the authors suggest a two-stage treatment whereby the microorganisms are immobilized separately and each transition is performed at its optimum temperature. This solution will, however, bring the original problem back. Namely, the inactivation of L-sorbose dehydrogenase by L-sorbosone which accumulates in the gel matrix.

Different stabilities of two biocatalytic species in a co-immobilized system will be found in several examples to be discussed below. One possible solution to this general problem could be the utilization of thermophilic organisms[15] and thermally stable enzymes[16] isolated from these thermophilic organisms.

Photobiological production of hydrogen involves several biocatalytic species which interact in a commensal way. Co-immobilization has therefore been used in order to improve productivity and stability of these production systems.

The combinations of co-immobilized biocatalysts for use in hydrogen production involve two microorganisms: *Chlorella vulgaris* and *Clostridium butyricum* as well as *Rhodospirillum rubrium* and *Klebsiella pneumonie:* an organelle and a microorganism: spinach chloroplasts and *Clostridium butyricum;* an enzyme and an organelle: hydrogenase from *Desulfovibrio gigas* and spinach chloroplasts. The different examples will be discussed below in reference to which type of biocatalyst has been used.

The photobiological production of hydrogen from water using spinach chloroplasts and a hydrogenase was first demonstrated by Benemann et al.[17] The isolated chloroplasts, however, had a very limited life time and the hydrogenase was inactivated by the oxygen produced. The latter problem was overcome to some extent by adding glucose oxidase to the system.

In an attempt to design a more stable system, *Chlorella vulgaris* and *Clostridium butyricum* were co-immobilized in agar gel with NADP as an electron carrier for continuous hydrogen evolution from water.[18] The principle for hydrogen evolution using co-immobilized *C. vulgaris* and *C. butyricum* is shown in Figure 2. As mentioned above this is a commensual mixed culture, involving two microorganisms and one external co-factor, NADP. Photosystem I and II in *C. vulgaris* reduce NADP and hydrogen is evolved through oxidation of NADPH with the hydrogenase from *C. butyricum*.

The optimum pH for the system was determined to be 7.0 and the optimum temperature 37°C, which actually is seven degrees above the optimum temperature for immobilized *C. vulgaris.* Nevertheless, the co-immobilized system evolved hydrogen at a constant rate for 6 days, indicating that the amount of *C. vulgaris* co-immobilized was well in excess of what was needed to balance the other components. As a matter of fact, no saturation concentration was found for either NADP or *C. butyricum*. Furthermore, the amount of *C. butyricum* used in the co-immobilized preparation was an order of magnitude lower than the maximum amount reported in the paper. The amount of *C. vulgaris* therefore is well in excess of what should be needed to match the quantities of the other two components, which might explain the apparent temperature stability of *C. vulgaris* in the co-immobilized system.

In view of this, the author's conclusion that the rate limiting step of the system is the diffusion of NADPH into the hydrogenase system also seems doubtful, since the system has not been supplied with a saturating amount of *C. butyricum*. Furthermore, as the report does not include a comparison between the co-immobilized system and the free system, it is not possible to conclude whether or not the proposed diffusion limitation occurs across the cell wall of the bacteria or is due to the gel in which the cells are immobilized. Most likely it is the latter case.

It seems to be a prerequisite, however, to use immobilized biocatalysts for this system in order to stabilize the hydrogenase of the bacteria against the oxygen evolved by the algae. The authors were previously able to demonstrate that the hydrogenase system of viable hydrogen producing bacteria could be stabilized by immobilization.[19]

Hydrogen evolution has also been studied in another system, consisting of two co-immobilized microorganisms, *Rhodospirillum rubrium* and *Klebsiella pneumoniae* in agar.[20] This example might be regarded as a curiosity since the co-immobilization was not initially planned, and was a result of a contamination of the *R. rubrium* culture with *K. pneumoniae.* This co-immobilized system was able to utilize glucose as well as acid hydrolyzed cellulose and wood sawdust as substrates for hydrogen production. *K. pneumoniae* was additionally able to produce hydrogen because of its ability to utilize dextrose and other oligosaccharides from which it produced organic acids. These acids were subsequently utilized by *R. rubrium*. The authors are very optimistic

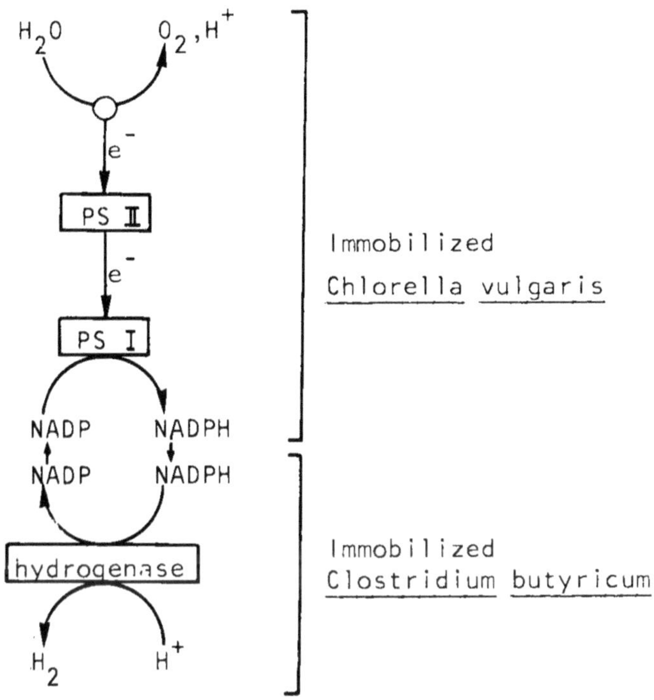

FIGURE 2. Diagramatic representation of hydrogen evolution by the immobilized *Chlorella vulgaris-Clostridium butyricum* system. (From Kayano, H., et al., *Biochem. Biophys. Acta*, 638, 80, 1981. With permission.)

about their system, which theoretically might produce 6 mol of hydrogen per mol of dextrose as compared to 2 mol of hydrogen per mol of glucose when using *Clostridium butyricum* alone.[21] However, it remains to be established whether or not co-immobilizing the separately grown cultures of *R. rubrium* and *K. pneumoniae* really can produce more hydrogen from dextrose than either organism does separately.

Light induced *in situ* oxygen production in algae has been successfully demonstrated using a commensal type arrangement of aerobic bacteria co-immobilized with algae. Two reports have recently appeared.

Wikström et al.[22] reported on the production of α-ketoacids using *Providencia sp.*, PCM 1298 co-immobilized with *Chlorella vulgaris*, UTEX 259. In this system the bacteria contributes amino acid oxidase activity and the algae, when illuminated, supplies the oxygen necessary for the reaction. The authors report a threefold increase in productivity when using illuminated algae as compared to an earlier, not yet published system where activated charcoal had been used to hydrolyze the hydrogen peroxide produced in the reaction.

The work also includes a study in the optimal ratio between algae and bacteria. No optimum is reported. The authors interpret their data in terms of a saturation level around a ratio of 2, which they attribute to diffusion limitations either of substrate or light. This leads to the fundamental question of whether or not an immobilized system should be used in this application. There are no data in this report, or in a previous report using a similar system,[23] to support the use of an immobilized system rather than a free system.

An attempt was made by Wikström et al.[22] to study their co-immobilized system under continuous operation over a 2-week time period. The preparation was studied for about 1 day each time at the beginning and at the end of this interval. In between measurements, the preparation was stored in the cold. It was found that 70% of the activity was retained after storage.

The authors attribute the decrease in overall activity to a decrease either in amino acid oxidase activity in the bacteria to loss of photosynthetic capacity of the algae or both. The reason for the decrease in activity is not analyzed further, however, one possibility might be the accumulation of hydrogen peroxide within the immobilization matrix. In previous reports the same group has used manganese oxide to hydrolyze the hydrogen peroxide and at the same time supply the system with oxygen. The co-immobilization of manganese oxide, at least theoretically, seems to be a more suitable way to create *in situ* oxygenation for this particular system, however, long-term stability data are necessary to validate this hypothesis.

The other study involving co-immobilized algae *Chlorella pyrenoidosa*, for *in situ* oxygen supply involves the production of dihydroxyaceton from glycerol by *Gluconobacter oxydans*.[24] It has been previously suggested that if the problem of *in situ* oxygenation of immobilized glycerol oxidizing bacteria could be solved, a continuous process for dihydroxyacetone production using immobilized bacteria would be industrially feasible.[25] Adlercreutz and co-workers have used the *Gluconobacter oxydans* to convert glycerol to dihydroxyacetone as an indicator reaction for the study of *in situ* oxygenation using co-immobilized algae.

Their results yield optimum values for pH, carbonate concentration, light intensity, concentration of bacteria, and algae. However, the most important result with reference to immobilization is that the production rate increases about 30% when bacteria and algae are co-immobilized within the same matrix as compared to immobilization of bacteria and algae in separate matrices.

It is absolutely necessary to obtain this kind of information when dealing with co-immobilization of two biocatalytic species especially in cases where it is reasonable to assume that one species will lose activity before the other. In such a case separately immobilized biocatalysts are preferable unless the co-immobilization, as shown above, actually offers an advantage.

The production rate of the co-immobilized system of *G. oxydans* and *C. pyrenoidosa* was further studied as a function of storage time. Only a slight decrease in activity was found over a 2-week period.

IV. ORGANELLES CO-IMMOBILIZED WITH MICROORGANISMS

The objective of producing hydrogen by biophotolysis of water has also been approached by using spinach chloroplasts co-immobilized with *Clostridium butyricum* by the same group that studied the co-immobilized system of *C. vulgaris* and *C. butyricum*.[26]

Agar was the carrier gel and ferridoxin or bensylviologen were used as electron carriers. The principle for hydrogen evolution by chloroplasts co-immobilized with *C. butyricum* is shown in Figure 3. The optimum pH for both catalysts was found to be 8.0, whereas the optimum temperature was widely different: 25°C for the chloroplasts and 37°C for the hydrogenase activity of *C. butryricum*. 30°C was chosen for the coupled reaction.

This co-immobilized system evolved hydrogen at a rate of 1.15 μmol/h × mg Chl, which compares well with the system composed of co-immobilized *C. vulgaris* and *C. butyricum*, where a rate of 0.29 to 1.34 μmol/h × mg Chl was measured. However, the immobilized chloroplasts have a very limited stability in that they lose 30% of their initial activity already after 6 hr of reaction time. The report does not clarify if this is due to the fact that the co-immobilized system was operated five degrees above the

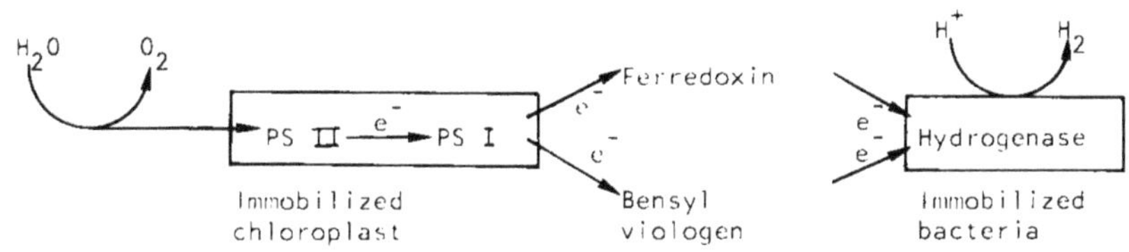

FIGURE 3. Diagramatic representation of hydrogen evolution by an immobilized chloroplast - *Clostridium butyricum* system. (From Karube, I., et al., *Biochem. Biophys. Acta*, 637, 490, 1981. With permission.)

optimum temperature. The authors indicate, rather, that instability is an inherent quality of isolated chloroplasts and they report that immobilization as such, improves their stability about three times.

Another drawback of the co-immobilized chloroplast-bacteria system is that bensylviologen is not stable in the presence of oxygen. As such it cannot be used for many cycles of oxidation-reduction. On the other hand, bensylviologen has a lower molecular weight than ferridoxin and it therefore more easily permeates the gel matrix and the cell walls of the bacteria.

Comparing the two co-immobilized systems developed for hydrogen evolution, *C. vulgaris* co-immobilized with *C. butyricum* at this stage of development seems to be the most successful.

V. ENZYMES CO-IMMOBILIZED WITH MICROORGANISMS

The co-immobilization of one or more enzymes with a microorganism has been discussed in several reports. The microorganisms have been used either as a source of one single enzymatic step or as a source of an entire metabolic pathway as will be discussed in the example below.

A bioselective membrane electrode that measures NAD^+ was constructed by co-immobilizing *Escherichia coli* ATCC 27195 and NADase (E.C.3.2.2.5) from *Neurospora crassa* between the gas permeable membrane of an ammonia sensor and a dialysis membrane (Figure 4).[27] The reaction scheme of this co-immobilized electrode is diagramed in Figure 4. As can be seen, the process is a commensural reaction involving two enzymes, one as an extract and one in intact bacterial cells.

The lifetime of the co-immobilized enzyme-bacteria electrode was at least 1 week and the authors verify that both catalytic species were necessary to enable the electrode to respond to NAD^+. The electrode gave a linear response in the range of 2.5×10^{-4} to 2.5×10^{-3} M NAD^+.

The main limitation for the use of this electrode seems to be its lack of selectivity, since the response curve for glutamine is similar to that of nicotine amide. This result seems to be due to the presence of a glutamine deaminase in *E. coli*. The authors suggest the use of a specific glutamine deaminase inhibitor, rather than purifying the nicotine amide deamidase from *E. coli* to overcome this problem. The enzyme does not seem to be co-factor dependent but might have a limited stability outside the cell.

The large-scale technical exploitation of immobilized enzymes has so far been limited by the fact that most enzymes of interest have been co-factor dependent. A great many research reports deal with the problem of co-factor regeneration. So far, this problem has been solved in an academic way rather than by finding solutions that can be applied in large-scale technical processes. However, a recent report presents a solution that could be used in technical applications. Burstein et al.[28] report on the co-immobilization of alcohol dehydrogenase and *E. coli* by glutaraldehyde in the presence of serum albumin.

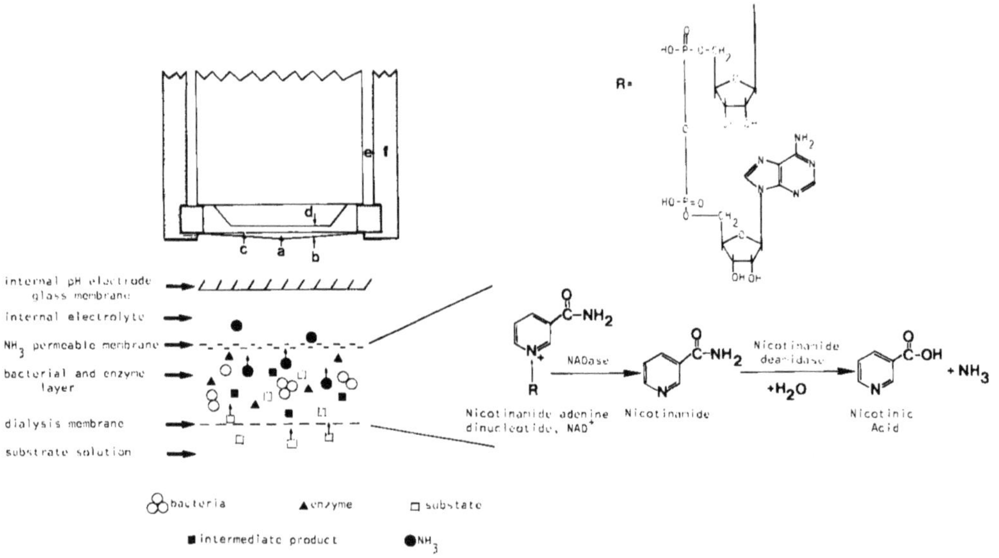

FIGURE 4. Schematic diagram and reaction scheme for a hybrid electrode used for NAD⁺ measurements. (a) Bacterial and enzyme layer, (b) dialysis membrane, (c) gas-permeable membrane, (d) internal sensing element, (e) internal filling solution, (f) electrode body. (From Riechel, T. L. and Rechnitz, G. A., *J. Membrane Sci.*, 4, 243, 1978. With permission.)

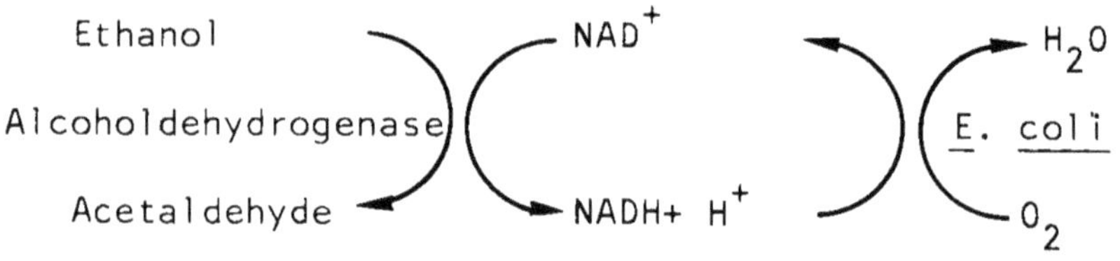

FIGURE 5. Diagramatic representation for NAD⁺ regeneration using alcohol dehydrogenase and intact cells of *E. coli*.

The respiratory chain of the bacteria can regenerate NAD⁺ from NADH without purification of the NADH oxidase. This regeneration step requires only oxygen. With excess oxygen present in the system it was possible to regenerate NAD⁺ at least 100 times (Figure 5).

Immobilization of the intact bacterial cells has one further advantage in that the respiratory chain becomes less sensitive to denaturation with an organic solvent, methanol. This opens up the possibility of utilizing water insoluble substrates such as steroids.

The co-immobilization of an enzyme and a microorganism has also been used in order to make nonmetabolizable substrates available for microorganisms.[29] This general approach was first demonstrated by fermenting cellobiose with *Saccharomyces cerevisiae* co-immobilized with β-glucosidase and fermentation of whey permeate with *Saccharomyces cerevisiae* co-immobilized with β-galactosidase. In this co-immobilization the enzyme was covalently bound to alginate by a carbodiimide coupling. The enzyme-alginate complex was then co-entrapped with yeast in alginate (Figure 6).

Naturally occurring cellulolytic enzyme systems are generally deficient in β-glucosidase activity. Therefore cellulose hydrolysates obtained from enzymatically saccharified cellulose are always composed of a mixture of glucose and cellobiose. Cellobiose can only be fermented by anaerobic bacteria, which means that apart from ethanol organic acids are produced, which decrease the product yield.

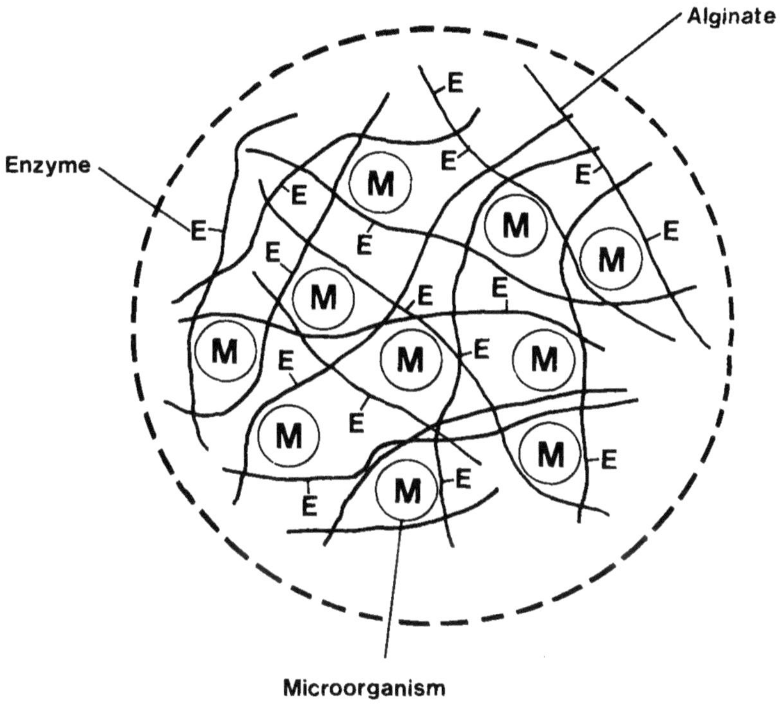

FIGURE 6. Diagramatic representation for co-immobilizing an enzyme and a microorganism within calcium alginate.

An artificial cellulose hydrolysate composed of a mixture of 5% (w/v) cellobiose and 5% (w/v) glucose could be continuously fermented to theoretical yields of ethanol using *S. cerevisiae* co-immobilized with β-glucosidase from sweet almonds[30] at dilution rates of 0.2 hr^{-1} or lower. The system operated continuously for 3 weeks with unaltered productivity at room temperature. A better productivity was obtained when the co-immobilized system was operated at higher temperatures. However, the β-glucosidase from sweet almonds has a very poor thermal stability, which makes the operational stability of the system very limited (unpublished data).

The effect of co-immobilizing both catalytic species in one carrier as compared to separate immobilization was compared in batch experiments (unpublished results). The production of ethanol from cellobiose was 44 μg ethanol/min in the co-immobilized system as compared to 28 μg ethanol/min when the enzyme and the yeast were immobilized separately. However, when either of the biocatalytic species was free the rate was faster than in the co-immobilized system. This clearly demonstrates the diffusion restrictions imposed by the gel matrix. However, the rationale for co-immobilizing β-glucosidase and yeast for fermenting cellobiose to ethanol was to create a metabolic pathway by which cellobiose is completely converted to ethanol without byproduct formation. Immobilization facilitates the recovery of biocatalysts which is crucial for the economy in a process for converting cellulose to ethanol in which the above sequence is the last step.

Instead of first covalently binding an enzyme to alginate, and then co-immobilizing it with a microorganism in alginate, the co-immobilization of amyloglucosidase and yeast was performed by covalently binding the enzyme to Dextran T 500.[31] This co-immobilized preparation was successfully used for brewing low calorie beer. Table 1 shows a comparison of crucial parameters for conventionally prepared low calorie beer and beer produced with the co-immobilized preparation. It appears that the co-immo-

Table 1
PREPARATION OF LOW CALORIE BEER WITH CO-IMMOBILIZED YEAST AND AMYLOGLUCOSIDASE

	Beer from reactor	Conventionally prepared beer
Original extract % P	8.21	8.0
Degree of fermentation %	86	86
Alcohol % w/w	3.56	3.40
Acetaldehyde ppm	5.1	9.8
Dimethyl sulfide ppm	0.046	0.092
Propionaldehyde ppm	0.18	+
Ethyl formate ppm	1.8	2.0
Ethyl acetate ppm	23	19
1-Propanol ppm	11	17
2-Methyl-2-Propanol ppm	8.3	8.2
3-Methyl-Butyl acetate ppm	0.47	2.0
1-Butanol ppm	+	+
3-Methyl-2-Butanol ppm	45	78
Ethyl hexanoate ppm	0.11	0.19
Ethyl octanoate ppm	0.11	0.20
Diacetyl	0.14	0.10

From Godtfredsen, S. E., et al., *EBC Congress 1981*, 505, 1981. With permission.

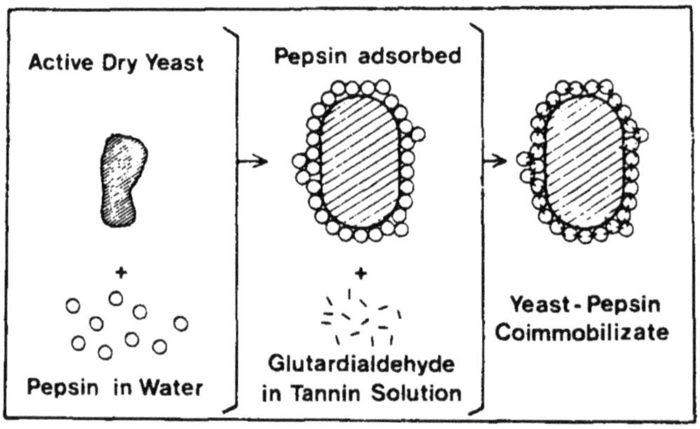

FIGURE 7. Diagramatic representation for co-immobilizing pepsin and yeast using glutardialdehyde. (From Hartmeier, W., *Current Developments in Yeast Research*, Pergamon Press, Canada, 1981, 105. With permission.)

bilized biocatalytic preparation produces a comparable quality of low calorie beer. The reactor containing co-immobilized amyloglucosidase and yeast was in operation for more than 1 month and the authors are optimistic about its future applications.

The idea of co-immobilizing an enzyme and a microorganism was recently approached in a different way.[32,33] The enzyme is first allowed to adsorb to the yeast cell, after which it is fixed on the surface by cross-linking with glutardialdehyde in tannin solution (Figure 7). This technique was applied to pepsin co-immobilized with *Saccharomyces cerevisiae* and cellulase also co-immobilized with *S. cerevisiae*.

The pepsin preparation was used for German white wine must production. 90% less foam formation and a somewhat increased fermentation rate was observed. The amino

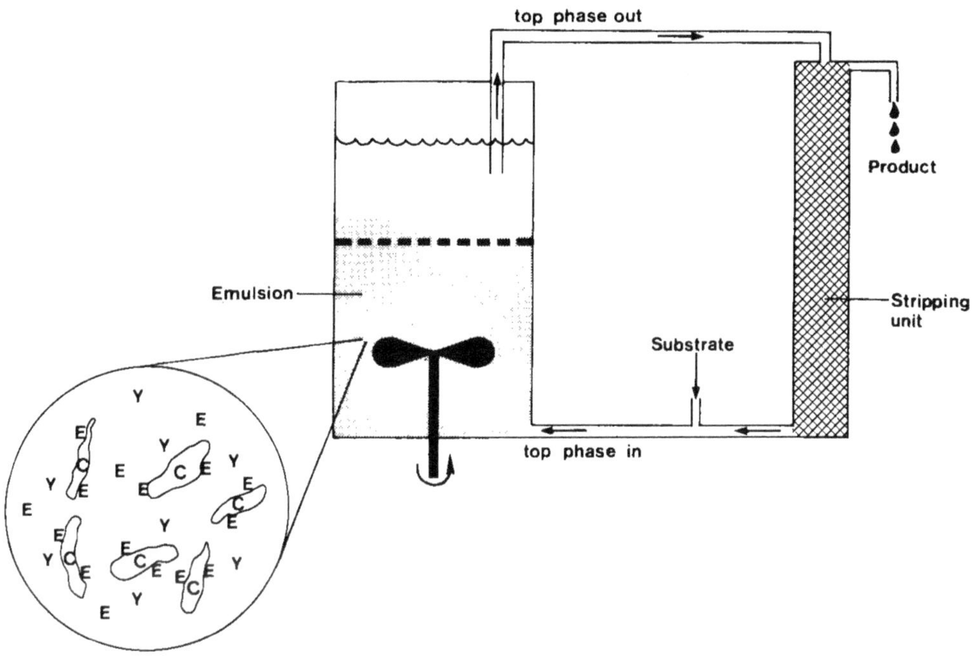

FIGURE 8. Diagramatic representation for continuous conversion of a particulate substrate using soluble temporarily co-immobilized enzymes and microorganisms.

acids liberated by the pepsin action were postulated to serve as nutrients for the yeast, thereby resulting in a higher fermentation rate. The author also mentions the use of soluble pepsin, however, it was found to be less efficient than the co-immobilized preparation. No data for this parameter are presented in the paper.

The cellulase preparation was claimed to be used for the conversion of cellulosic material to ethanol. However, as in the previously described work[30] it was only used for converting cellobiose to ethanol. Due to steric hindrance, it would not be possible to degrade a macromolecular substrate with cross-linked enzymes. It therefore seems questionable to immobilize a cellulase preparation on the surface of the yeast cells. Higher enzymatic activity might have been achieved if a β-glucosidase preparation had been used instead. The paper describes the amount of co-immobilized preparation needed for complete concersion of 1% cellobiose to ethanol but it is not discussed whether the enzyme or the yeast is the rate limiting factor.

The idea of direct conversion of cellulose to ethanol put forward in the previous paper was demonstrated using a different immobilization technique.[34] The cellulolytic enzyme system from *Trichodema viride* was co-immobilized with *S. cerevisiae* in an aqueous two-phase system. The biocatalysts remain soluble and are temporarily immobilized (Figure 8). The technique is described in more detail elsewhere in this volume. Due to the fact that the biocatalysts are soluble macromolecular and solid substrates like cellulose can be converted to the endproduct, ethanol. The large surface area created between the phases also facilitates the diffusion of product from the biocatalysts, thereby increasing the productivity of the system.

VI. ENZYMES CO-IMMOBILIZED WITH ORGANELLES

Continuous hydrogen evolution from water was demonstrated in yet another co-immobilized system.[35] In contrast to the previously described systems, the biocatalytic species were isolated and co-immobilized.

Chloroplasts and Sepharose-bound hydrogenase from *Desulfovibrio gigas* were co-entrapped in alginate and cytochrome c_3, ferridoxin and flavodoxin were used as electron carriers. Ferridoxin and flavodoxin were better electron acceptors than cytochrome c_3. Both spinach and lettuce chloroplasts were studied and it was found that spinach chloroplasts were more effective for hydrogen photo production.

The sensitivity of hydrogenase to oxygen denaturation was overcome by immobilizing the enzyme onto Spherosil and Sepharose 4B. The enzyme did not lose any activity after 19 hr of oxygen exposure.

In contrast to the previously described co-immobilized systems for hydrogen production this report clearly shows the advantage of co-immobilization. A fivefold increase in hydrogen production is obtained when the immobilized enzyme is co-entrapped with chloroplasts as compared to a system where the two catalytic species are immobilized separately.

Thus, this report demonstrates two major advantages of immobilization: first, the protection against oxygen denaturation; second, the increase in the overall activity due to the close proximity achieved by co-entrapment. The major disadvantage of this co-immobilized system is again the limited stability of the chloroplasts.

VII. CONCLUSIONS

From this literature review it appears that even though co-immobilization of biocatalysts such as enzymes, organelles, and microorganisms is still a technically rather unexplored field, the current research in this area points out that it has a great potential for the future. By co-immobilizing different biocatalytic species, new metabolic pathways are created and this technique therefore is a valuable complement to genetic manipulation of microorganisms.

The main problem in technical exploitation of co-immobilized systems as described above is the difference in stability between the two catalytic species. The difference in thermal stability especially restricts the productivity of the overall system as well as the lifetime of the co-immobilized system, emphasizing again that "the chain is not stronger than the weakest link". By using new immobilization techniques such as aqueous two-phase systems and membrane reactors, discussed elsewhere in this volume, some of these problems could, however, be overcome. Using these techniques, the biocatalysts remain soluble even when immobilized. It is possible, as well, to add a catalytic species to the system when its activity is declining without interrupting ongoing processes. This means that every catalytic species in a co-immobilized system of this kind will be optimally utilized. When, however, stable biocatalysts are utilized, co-immobilization in solid supports should have a great bearing in a number of future biotechnical applications.

REFERENCES

1. **Harrison, D. E. F.,** Mixed cultures in industrial fermentation processes, *Adv. Appl. Microbiol.,* 24, 129, 1978.
2. **Meers, J. L.,** Growth of bacteria in mixed cultures, *Crit. Rev. Microbiol.,* 2, 139, 1973.
3. **Mattiasson, B.,** Biochemical applications and perspectives of immobilized multistep enzyme systems, in *Biomedical Applications of Immobilized Enzymes and Proteins,* Chang, T. M. S., Ed., Plenum Press, New York, 1977, 253.
4. **Young, J. C. and McCarty, P. L.,** The anaerobic filter for waste treatment, *J. Water Poll. Control Fed.,* 41(5,2), 160, 1969.

5. Switzenbaum, M. S. and Jewell, W. J., Anaerobic attached-film expanded-bed reactor treatment, *J. WPCF*, 52(7), 1953, 1980.
6. Frostell, B., Wastewater treatment and energy recovery in an expanded bed system, in 3rd International Congress on Industrial Waste Water and Wastes, Stockholm, February 6-8, 1980.
7. Bryant, M. P., Microbial methane production — theoretical aspects, *J. Animal Sci.*, 48(1), 193, 1979.
8. Ljungdahl, L. G., Carreira, L., and Wiegel, J., Production of ethanol from carbohydrates using anaerobic thermophilic bacteria, *The Ekman-Days 1981*, 4, IV:23-28, 1981.
9. Lettinga, G., van Velsen, A. F. M., de Zeeuw, W., and Hobma, S. W., The First International Symp. on Anaerobic Digestion, 17th-21st Sept., 1979, University College, Cardiff.
10. Christensen, M. H. and Haremoes, P., Biological denitrification of sewage: a literature review, *Prog. Wat. Tech.*, 8(4/5), 509, 1977.
11. Karube, I., Kuriyama, S., Matsunaga, T., and Suzuki, S., Methane production from wastewaters by immobilized methanogenic bacteria, *Biotechnol. Bioeng.*, 22, 847, 1980.
12. Karube, I., Mitsuda, S., Matsunaga, T., and Suzuki, S., A rapid method for estimation of BOD by using immobilized microbial cells, *J. Ferment. Technol.*, 55(3), 243, 1977.
13. Compere, A. L. and Griffith, W. L., Fermentation of waste materials to produce industrial intermediates, *Dev. Ind. Microbiol.*, 17, 247, 1976.
14. Martin, C. K. A. and Perlman, D., Conversion of L-sorbose to 2-keto-L-gulonic acid by mixtures of immobilized cells of *Gluconobacter melanogenus* IFO 3293 and *Pseudomonas* species, *Eur. J. Appl. Microbiol.*, 3, 91, 1975.
15. Ljungdahl, L. G., Physiology of thermophilic bacteria, *Adv. Microbial Physiol.*, 19, 149, 1979.
16. Mitchell, R. W., Hahn-Hägerdal, B., Ferchak, J. D., and Pye, E. K., Characterization of β-glucosidase activity in *Thermoanaerobacter ethanolicus*, 4th Symposium on Biotechnology in Energy Production and Conservation, Gatlinburg, TN, May 11-14, 1982.
17. Benemann, J. R., Berenson, J. A., Kaplan, N. O., and Kamen, M. D., Hydrogen evolution by a chloroplast-ferredoxin-hydrogenase system, *Proc. Natl. Acad. Sci.*, 70(8), 2317, 1973.
18. Kayano, H., Matsunaga, T., Karube, I., and Suzuki, S., Hydrogen evolution by co-immobilized *Chlorella vulgaris* and *Clostridium butyricum* cells, *Biochem. Biophys. Acta*, 638, 80, 1981.
19. Suzuki, S., Karube, I., and Matsunaga, T., Application of a biochemical fuel cell to waste waters, *Biotechnol. Bioeng. Symp.*, 8, 501, 1978.
20. Weetall, H. H., Sharma, B. P., and Detar, C. C., Photometabolic production of hydrogen from organic substrates by free and immobilized mixed cultures of *Rhodospirillum rubrium* and *Klebsiella pneumoniae*, *Biotechnol. Bioeng.*, 23, 605, 1981.
21. Karube, I., Matsunaga, T., Tsuru, S., and Suzuki, S., Continuous hydrogen production by immobilized whole cells of *Chlostridium butyricum*, *Biochem. Biophys. Acta*, 444, 338, 1976.
22. Wikstrom, P., Szwajcer, E., Brodelius, P., Nilsson, K., and Mosbach, K., Formation of α-keto acids from amino acids using immobilized bacteria and algae, *Biotechnol. Lett.*, 4(3), 153, 1982.
23. Brodelius, P., Nilsson, K., and Mosbach, K., Production of α-keto acids. Part I. Immobilized cells of *Trigonopsis variabilis* containing D-amino acid oxidase, *Appl. Biochem. Biotechnol.*, 6, 293, 1981.
24. Adlercreutz, P., Holst, O., and Mattiasson, B., Oxygen supply to immobilized cells. II. Studies on a co-immobilized algae-bacteria preparation with *in situ* oxygen generation, *Enzyme Microb. Technol.*, 4, 395, 1982.
25. Nabe, K., Izuo, N., Yamada, S., and Chibata, I., Conversion of glycerol to dihydroxyacetone by immobilized whole cells of *Acetobacter xylinum*, *Appl. Envir. Microbiol.*, 38(6), 1056, 1979.
26. Karube, I., Matsunaga, T., Otsuka, T., Kayano, H., and Suzuki, S., Hydrogen evolution by co-immobilized chloroplasts and *Clostridium butyricum*, *Biochem. Biophys. Acta*, 637, 490, 1981.
27. Riechel, T. L. and Rechnitz, G. A., Hybrid bacterial and enzyme membrane electrode with nicotinamide adenine dinucleotide response, *J. Membrane Sci.*, 4, 243, 1978.
28. Burstein, C., Ounissi, H., Legoy, M. D., Gellf, G., and Thomas, D., Recycling or NAD^+ using coimmobilized alcohol dehydrogenase and *E. coli*, *Appl. Biochem. Biotechnol.*, 6, 329, 1981.
29. Hägerdal, B., Enzymes co-immobilized with microorganisms for the microbial conversion of non-metabolizable substrates, *Acta Chem. Scand.*, B 34(8), 611, 1981.
30. Hägerdal, B., Lopez-Leiva, M., and Mattiasson, B., Membrane technology applied to bioconversion of macromolecular substrates and upgrading of products. A model study on ethanol production from cellulose, *Desalination*, 35, 365, 1980.
31. Godtfredsen, S. E., Ottesen, M., and Svensen, B., Application of immobilized yeast and yeast coimmobilized with amyloglucosidase in the brewing process, *EBC Congress 1981*, 505, 1981.
32. Hartmeier, W., *Saccharomyces cerevisiae* coimmobilized with pepsin for wine making, in *Current Developments in Yeast Research*, Stewart, G. G. and Russell, I., Eds., Pergamon Press, Canada, 1981, 105.
33. Hartmeier, W., Basic trials on the conversion of cellulosic material to ethanol using yeast coimmobilized with cellulolytic enzymes, in *Adv. in Biotechnology*, Vol. 3, Moo-Young, M., Gen. Ed., Vezina, C. and Singh, K., Eds., Pergamon Press, Canada, 1981, 377.

34. **Hahn-Hägerdal, B., Mattiasson, B., and Albertsson, P.-Å.**, Extractive bioconversion in aqueous two-phase systems. A model study on the conversion of cellulose to ethanol, *Biotechnol. Lett.*, 3(2), 53, 1981.
35. **Cocquempot, M. F., Aguirre, E., Lissolo, T., Monsan, P., Hatchikian, E. C., and Thomas, D.**, Co-immobilization effect of H_2 production by a chloroplast membranes — hydrogenase system, *Biotechnol. Lett.*, 4(5), 313, 1982.

Chapter 6

ANALYTICAL APPLICATIONS OF IMMOBILIZED CELLS

Bo Mattiasson

TABLE OF CONTENTS

I. INTRODUCTION

When designing an analytical procedure certain key factors must be taken into account, e.g., specificity, stability, quickness, and reproducibility. Biochemical analysis is satisfactory in some of these respects. When using enzymes in analysis, the specificity criteria are often met already by the inherent properties of the enzyme.[1,2] It has been more difficult to secure stability. However, with the introduction of immobilized enzymes a relatively high operational stability could be achieved[2] as well as the reuse of the biocatalyst. Solid phase technology has made it possible to develop both enzyme[3,4] and antibody-based[5,6] analytical reactions, some of which are commercially available as analytical kits or analytical instrumentations. As mentioned earlier (in Chapter 1, Volume I), immobilized cells can perform processes not easily done by immobilized enzymes. In theory, the immobilized cells should have a great potential in analysis, however, some drawbacks have been recognized. The most serious one being lack of specificity. Cells offer other possibilities that the enzymes cannot fulfill. Many toxic substances are toxic because they interfere with the biological membranes. All such physicochemical effects can be detected and quantified by immobilized cells or cell models. When using cells in analysis the same criteria can be set up as when designing large-scale processes based on immobilized cells. This means that living cells (Chapter 2, Volume II) as well as resting cells (Chapter 1, Volume II) and dead cells (Chapter 7, Volume II) can be used. The state of the cell used depends largely on its intended application. Analytical needs that can be satisfied by the use of immobilized cells are summarized in Table 1. Table 2 summarizes the different states of cells applied in analysis.

II. EXPERIMENTAL

A. Immobilization Methods

When selecting a method of immobilization one has to optimize different factors, as is discussed in greater detail in Chapter 2, Volume I.

First, the method of choice must result in a final preparation with cells still capable of performing the processes intended for the analytical step. This means that when the whole metabolism must be involved, or essential membrane processes are exploited the methods of immobilization chosen must be very gentle, whereas when only one or a few intracellular enzymes are expected to be used, also less gentle methods can be used.

In general, most systems designed so far have been based on enzyme sequences and thereby on the assumption that the major part of the metabolism is intact.

B. Immobilized Microbial Cell Preparations Used

Several different methods of immobilization have been used. When working with different enzyme electrodes the enzyme is commonly encaged around the sensitive tip of the electrode.[2] Usually the enzyme is kept in position by a semipermeable membrane covering the enzyme layer. The enzyme has been either free in solution, bound to small particles, or within membranes covering the sensitive tip of the transducer.[8] Each arrangement has its specific advantages and the same situation can be foreseen when cells are used instead of enzymes.

Free cells entrapped behind a semipermeable membrane are easier to use than enzymes. A common problem here is that the thickness of the layer between the electrode surface and the membrane must be reproducible when changing from one cell preparation to another, otherwise the cell electrode must be recalibrated for each cell preparation. Such calibration may be detrimental to cell-based systems, since many bioassays *per se* influence the total cell metabolism in such a way that the cell preparation

Table 1
ANALYTICAL PROBLEM

Analysis of:
- Specific substances
- Group of similar substances
- Groups of substances with similar effects on the cells/organelles
- Substances influencing the overall metabolism
- Specific agents capable of interacting with specific sites on the cells
- Overall measurements of total content of organic matter

Analysis involving:
- Coenzyme-dependent specific enzyme catalyzed reactions

Table 2
CONDITION OF THE CELLS USED

Only one or a few enzymes involved
Resting cells
Permeabilized cells
Living cells with an intact surface
Actively growing and dividing cells
Mixed cultures of living cells, or mixtures of enzymes and cells
Cell co-immobilized with a specific enzyme or organelle
Immobilized cells or organelles with a substantial part of the metabolism intact

is no longer useful for analysis. In such cases a simple system of disposable cell preparations is to be preferred.

Covalently bound cells are used but very little in biotechnology and have apparently never been used for analytical purposes.

1. Entrapped Cells

Cells entrapped within membranes have been used. From the early work of Broun[9] on cross-linked proteic membranes cell- and organelle-membranes have been developed that can be used for analytical purposes. Acrylic polymers may also be useful for entrapment, even if a substantial cell death in the polymerization step is to be foreseen.[10]

The carbohydrate polymers — carrageenan and alginate — used but little until recently are *per se* characterized by low mechanical stability, though secondary stabilization has improved the gels.[11] By casting the cell-containing gels on a nylon net backbone it proved possible to obtain alginate-membranes of good mechanical stability and with a high content of living cells.[12]

Many reports are available of immobilization of bacteria on an acetyl — cellulose filter with the use of agar. In a typical experiment[13] 20 mg of agar was dissolved in 900 $\mu\ell$ of physiological saline. After cooling to 53°C 100 $\mu\ell$ saline containing 10 mg of wet intact cells of *Leuconostoc mesenteroides* was added. The acetyl cellulose filter was dipped in the agar solution for 20 sec, cooled to 5 °C and was washed with physiological saline.

The use of a precolumn in front of the transducer, makes it possible to carry out all the immobilization procedures used when designing large-scale and laboratory operations. The essential point in this connection is that mechanical stability must be high and that cell leakage must be low, preferably nonexistent.

2. Biospecific Affinity Immobilization

Furthermore, besides these methods a biospecific reversible immobilization method may be used. Immobilized biomolecules with specificity for cell surface structures on the cell to be immobilized are packed in a column. The cells are introduced and trapped

Table 3
ASSOCIATION CONSTANTS FOR SOME NATURALLY OCCURRING REACTANT PAIRS

Reactant pair	Kass (l/mol)	Ref.
Avidin-biotin	10^{15}	18
Antibody-hapten	10^{5}—10^{11}	6
Protein A — Fc-region of IgG	10^{6}	19
Lectin — carbohydrate		
Simple sugars	10^{3}—10^{4}	20
Multipoint attachment	10^{6}—10^{7}	20

biospecifically.[14,15] In this way very gentle immobilization can be achieved without really influencing the cell-metabolism at all. The biospecific molecules used may be a broad range of structures with specificities for the surface of a certain cell-species or strain. So far, mainly lectins and antibodies have been used. Table 3 gives some of the biochemical reactant pairs that can be used and their association constants.

As pointed out earlier, it is essential that cell-populations used in bioassays can be easily replaced in a reproducible way. Most methods so far adopted involve a mechanical step when a piece of membrane, etc., is replaced by a fresh one. However, when using biospecificity reversible immobilization (= affinity binding) it has been shown that a chemical dissociation step may be sufficient and that after reconditioning of the adsorbent, cells can be introduced again. Enzyme immunoassays using such a repeated reaction cycle have been described as highly reproducible.[16,17]

This has been ascribed mainly to the fact that the experimental conditions were very stable and only a few parameters varied between the cycles and furthermore, that an excess of sorption molecules was used in the immobilized bed.[21] When using immunosorbents it is essential to have a complete washing step when regenerating the sorbent. When dealing with cells, this step is more difficult since the cells are immobilized by multipoint attachment thereby improving the strength of their attachment. Thus, when designing a fully reversible immobilization step for cells, biomolecules attaching themselves only loosely to the cells are to be preferred in order to facilitate the exchange of cells.

When the samples to be analyzed do not contain cells or other catalytic activities that may bind them to the sorbent, or when it does not matter whether unspecific adsorption occurs, also unspecific immobilization procedures may be used, e.g., ion exchange[22] or hydrophobic interactions.[23] Also spontaneous growth of microorganisms on the support may be utilized.[24]

C. Immobilized Mammalian Cells

Two main groups of cells can be distinguished, one requiring a support for growth (see Chapter 4, Volume I) and one growing in suspension. As mentioned in Chapter 4, Volume I, much work has been put into developing systems for culturing carrier-dependent cells. The results of that work can of course be utilized when designing analytical systems based on immobilized mammalian cells.

Furthermore, for suspension cultured cells, a biospecific reversible immobilization procedure may be adopted.[25] Since mammalian cells are much more sensitive than microbial cells, they were the most troublesome to immobilize in a viable state, but, on the other hand, such cell systems have a great potential when used in analytical systems. An example of the sensitivity of such cell preparations is the experiments by Meyerhoff[26] who tried to create a glutamine sensing electrode with the aid of a slice of porcine kidney. His efforts were successful as long as the cells were not treated with cross-linkers to stabilize them. Glutaraldehyde treatment reduced the activity to zero.

Mammalian cells have a great potential in analysis, but since the technology has so far been focused mainly on microbial cells, no tailor-made systems for these fragile animal cells are available. Thanks to their high sensitivity, mammalian cells are good models for quantifying unspecific physicochemical effects on cells and cell membranes. Owing to difficulties in obtaining fully reproducible systems, also simple cell models have been used.

D. Liposomes as Cell Models in Analysis

Leakage of substances through membranes has been used for free cells in ^{51}Cr-release test.[27] Owing to the lack of cell preparations of constant and high quality unless facilities for culturing cells are available, also cell models in the form of liposomes have been used.[28] These preparations, containing a demonstrable molecule, consist of a bilayer of membrane with essentially the same properties as living cells (Figure 1). Such liposomes were bound to porous filter paper by adsorption and the damaging effect on the membrane, measured as the degree of leakage of contents, was assessed and related to the amount of surface active substance present.

To improve the stability of the system, the liposomes were immobilized in Sepharose-gel with bound fatty acids. These hydrophobic gels interfere with the liposomes by immobilizing the preparation very firmly. Using such a preparation liposomes could be packed in columns and used for analysis in a continuous flow system.[30]

III. TRANSDUCERS USED

A crucial point when using immobilized species in analysis is the method of collecting and recording the signal. In conventional bioassays based on dividing cells or living whole organisms time consuming cell counting[31,32] or measurement of incorporated isotopes has often been used. In studies of macroscopic organisms their physical activity has been recorded by photo-cells and when measuring toxic levels of substances, LD_{50}-doses have been registered.

A. Electrodes

One of the aims in designing immobilized cell systems for bioassays was to develop simple and fast procedures to replace the above-mentioned methods. The combination of the sensor (the cells, organelles, or cell models) when placed close to the transducer, thereby enabling direct recording of metabolic events and thereby eliminating the need of the lag phase when a macroscopic change takes place that can be registered by microscopic observation or turbidimetric measurement. The transducer systems tried so far are probably only a few of all those potentially possible. The main systems so far used are given in Table 4. When designing such analytical systems large parts of the metabolism may be involved, but there are also examples when only a single enzyme of the cell is used. When it is difficult to obtain a good signal directly from the metabolic events, it is also possible to add exogenous mediators.[33] Several different electrode devices have been used (Table 4). The general principle is to cover the sensitive tip of the electrode with bioactive material — cells, organelles, tissue slices, etc., Figure 2 illustrates the preparation of a cell-electrode.

Polarographic oxygen electrodes have been used when a large part of the metabolic apparatus of the immobilized cells, including the electron transport system, is involved. Polarographic electrodes are easy to operate and give reliable results, but the response does not discriminate between different entries in the metabolic network, i.e., no specificity can be expected unless extraordinary arrangements are made. As discussed under

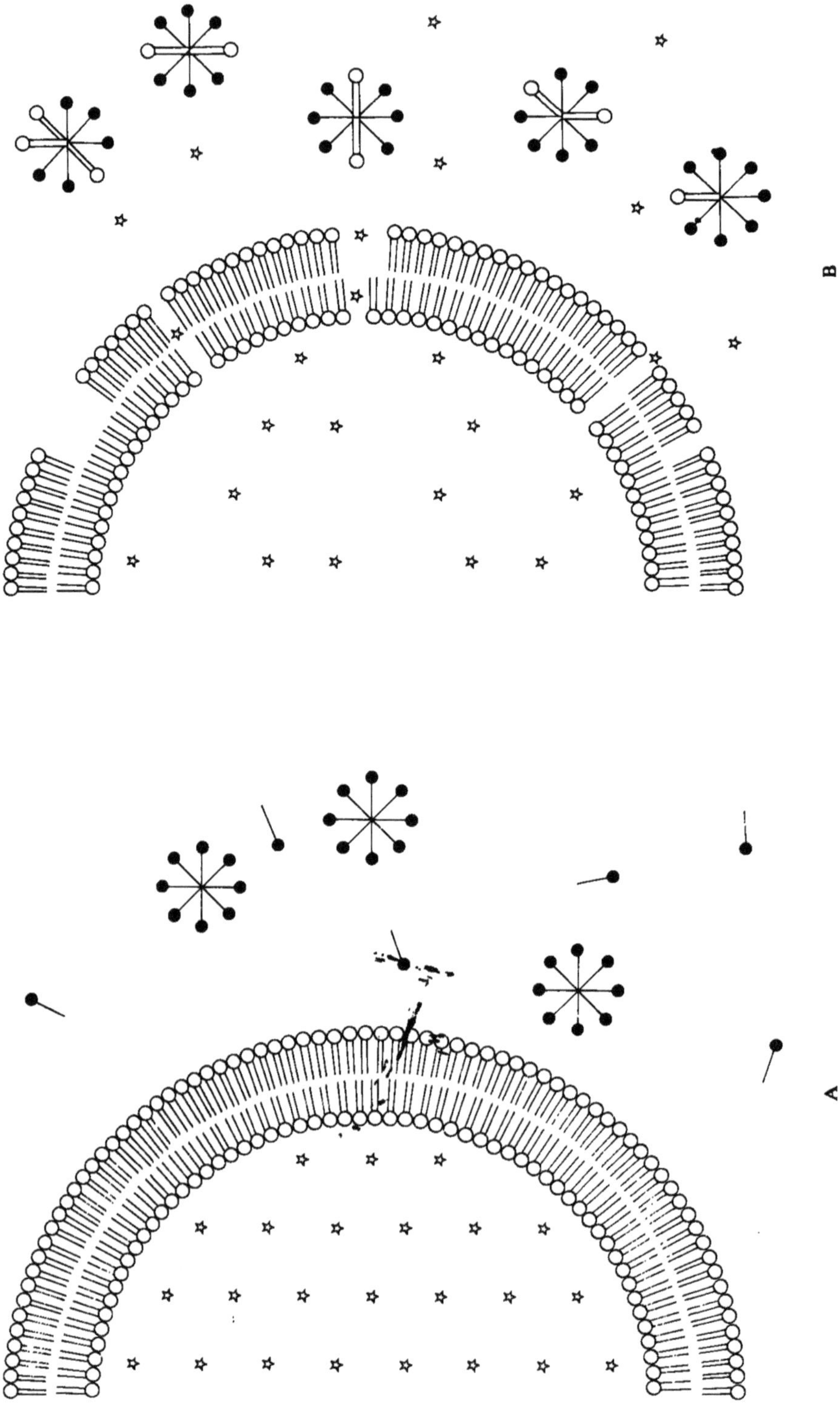

FIGURE 1. Schematic presentation of the effect of detergents on liposomes. Phospholipids (O=), peroxidase ✿ and detergent (●—). (A) Before release of enzyme the peroxidase is entrapped in liposomes and the detergent is either free or forming micelles. (B) After release of the enzyme, the phospholipids and detergent form mixed micelles and the peroxidase is liberated. (From Eriksson, H. and Mattiasson, B., *Appl. Biochem. Biotechnol.*, 7, 81, 1982. With permission.)

Table 4
TRANSDUCER USED TO CREATE BIOSENSORS

Ion selective electrodes
pO_2-electrodes
pCO_2-electrodes
Conductivity meter
Thermistor
Photometer
Mass spectrometer
Piezoelectric membranes

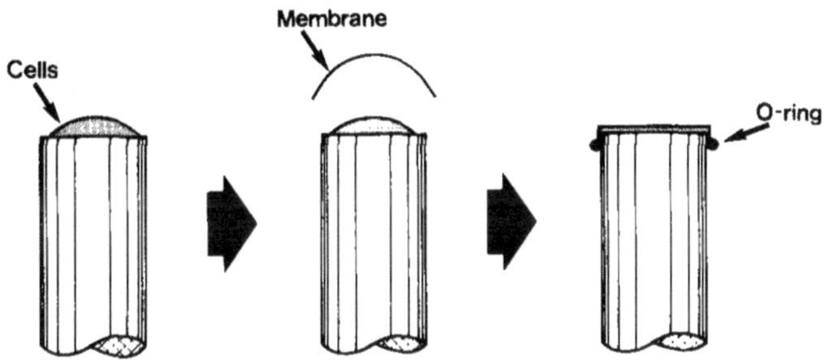

FIGURE 2. Schematic presentation on the procedure to prepare a cell-electrode.

"Operational conditions", there are some ways to improve the specificity. What is said for pO_2-electrodes in general holds also for pCO_2-electrodes.

Much work has been published on potentiometric NH_3-gas sensitive electrodes.[26,34-37] By using a selective electrode in which the sensitive part is protected from all substances in the sample except for gaseous compounds, a rather high selectivity can be achieved. As is discussed later, it may be difficult to discriminate between some metabolites when suitable NH_3-generating enzymes are present for more than one substrate in the immobilized cell preparation.

Conductivity meters are capable of recording changes in the medium when a neutral molecule, such as glucose is converted to charged products, such as lactic acid.[38] The transducers used so far can surely be improved. What was said about selectivity holds also for pO_2 and pCO_2-electrodes.

An even more common transducer is the thermistor/thermocouple/peltier element used in calorimetric devices. Here all reactions generating or consuming heat are registered and sometimes the signals obtained may be very complicated. On the other hand, such a general system has a great potential provided the selectivity of the biopart of the system is good. A simple flow-calorimeter called "microbe thermistor" has been used in such studies.[10] The apparatus is shown in Figure 3.

B. Photometer

The most widely used instrument in enzyme-based analysis has only rarely been used in combination with immobilized cells. This is due partly to its sensitivity to disturbances by a complex medium. The photometer will find its best applications in analyses where substances influencing the total cell metabolism are to be quantified, since in such assays a two-step incubation protocol is to be foreseen, and then the last step can be performed under well-defined conditions.[40,41]

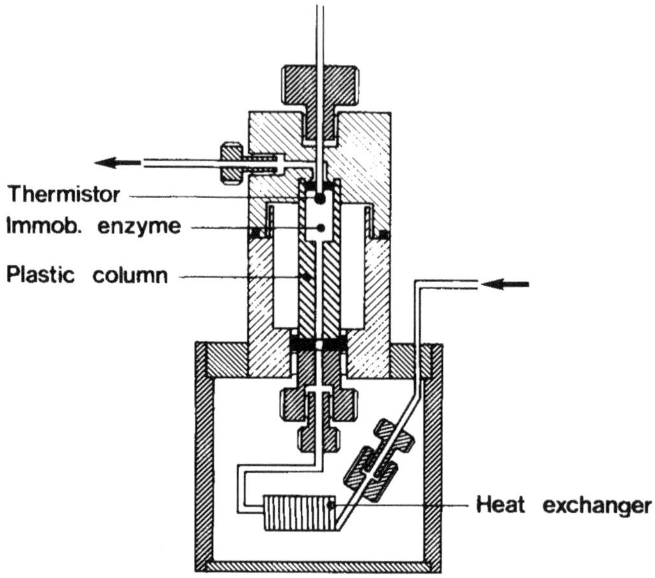

FIGURE 3. A single column enzyme thermistor. (From Mattiasson, B., et al., *Food Process Engineering*, Vol. 2, Linko, P. and Larinkari, J., Eds., Allied Science Publishers, London, 1980, 59. With permission.)

C. Mass Spectrometer

All volatile products from cell metabolism can be quantified with a mass spectrometer. The possibility of quantifying individual mass peaks enables quantification of individual substrates in complex media provided specific volatile intermediary metabolites are generated.[42,43] The experimental set-up is shown in Figure 4.

IV. EXPERIMENTAL PROCEDURE

A. Experimental Considerations

The paragraph on applications discusses two principally different ways of using the immobilized cell in analysis. When using only one or a few enzymes, the cell metabolism will recover and the cell electrode can then be used over and over again, whereas when the total metabolism is involved it will take the cells a long time to recover their original basic value of resting metabolic activity, if they ever do. In the latter case, the cell preparation has to be replaced by a fresh preparation before the next assay. Under such circumstances the cell preparations must be changed in a reproducible way.

A simple way is to use the cell preparation separated from the measuring device and expose only the incubation medium to the transducers or alternatively observe the events in the vessel containing the cell preparation without any direct contact between transducer and cell preparation.

In assays of the vitamin, nicotinic acid, Matsunaga et al.[45] used *Lactobacillus arabinosus* cells immobilized in agar. In each assay a constant amount of the immobilized cell preparation was added to the sample. After 1 h at 30°C the potential was measured. In this way separate samples can be incubated in sequence and assayed after a constant incubation time. With this procedure it is possible to analyze many samples in a relatively short time.

Another way to replace the cells was demonstrated by Mattiasson et al.[12] who used *Saccharomyces cerevisiae* entrapped in armed alginate membranes. The microbe mem-

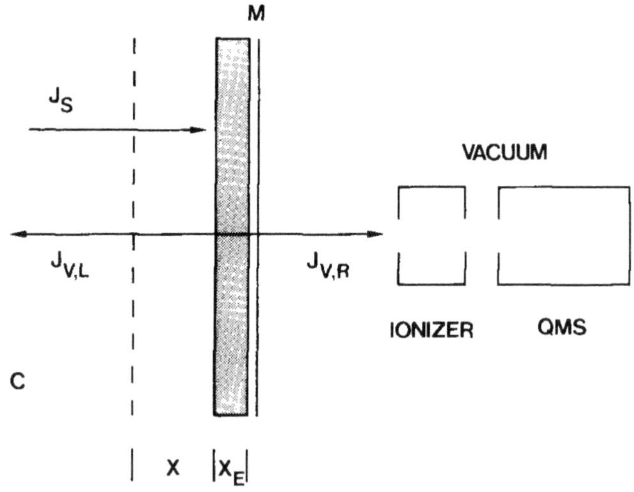

FIGURE 4. Basic configuration used for the mode that assays a nonvolatile substrate concentration, C, by means of a volatile product. M is the semipermeable membrane, with an enzyme layer of thickness X_E and an assumed aqueous unstirred layer of thickness X on the left. The vacuum of the mass spectrometer (here a quadrupole mass spectrometer (QMS) shown with its ionizer) is to the right of the membrane. J_S is the steady-state substrate flux to the enzyme layer, while $J_{v,R}$ and $J_{v,L}$ are the steady-state right- and left-going volatile product fluxes.[40] (From Weaver, J. C., *Biomedical Applications of Immobilized Enzymes and Proteins*, Vol. 2, Chang, T. M., Ed., Plenum Press, New York, 1980, 207. With permission.)

brane was cut into indentical pieces (6 mm diam.) and the assays were performed by incubating these standardized membranes in the samples to be analyzed. The membrane was then rinsed with buffer and placed on a pO_2-electrode. To arrange the membranes in a reproducible manner an electromagnet-induced magnetic field was used to secure a reproducible pressure on the membrane (Figure 5).

Another approach when the cells are immobilized by biospecific reversible immobilization[14] was adopted by Mattiasson and Johansson.[40] By using a disposable plastic syringe filled with a sorption material — immobilized lectins or antibodies — it was possible to immobilize cells in the different syringes in a reproducible way so that when incubating with substrate, the metabolic activity observed was reproducible. The metabolism was traced by either pH indicators in weakly buffered media[40] or by redox indicators in the medium (Figure 6).[41]

The group of assays in which the cell preparation has to be replaced after each assay embraces all those classical bioassays where a certain substance is quantified by its effect on the growth of a specific strain of dependent cell. As mentioned earlier, it might be possible to observe these differences much earlier than what was possible before. To improve the measurable response a large number of cells can be added to the sample to secure a marked metabolic potential each time. Immobilized cell preparations are such dense cell-populations. However, if cell division has to take place, too dense a cell preparation will inhibit the response since the cells will not be able to divide and thus the cell density will have to be optimized. An example of this is shown in Figure 7.[13]

A further area in which cells can be used in analysis is that in which physicochemical interferences of unspecific nature are to be quantified. Toxic effects on cell metabolism may be followed relatively easily, whereas unspecific influences on cell membranes are more difficult to quantitate. Here liposomes and other simple membrane models may be useful.

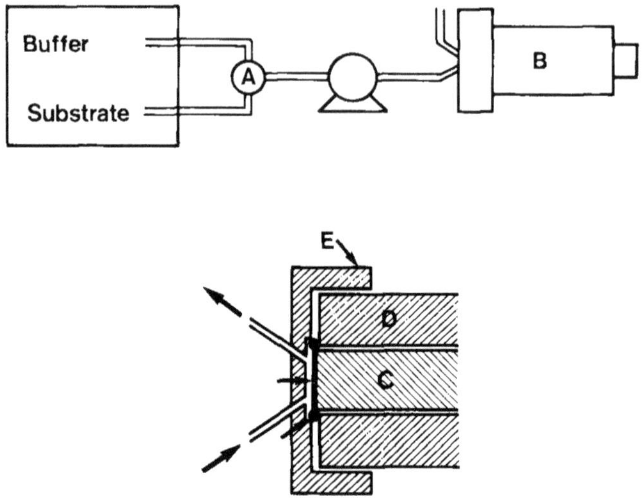

FIGURE 5. Experimental set-up when using a flow-through cell-electrode with an electromagnetic membrane holder. (A) Three-way valve, (B) electromagnetic membrane holder, (C) polarographic oxygen electrode, (D) electromagnet. (From Mattiasson, B., et al., *Enzyme Microb. Technol.*, 4, 153, 1982. With permission.)

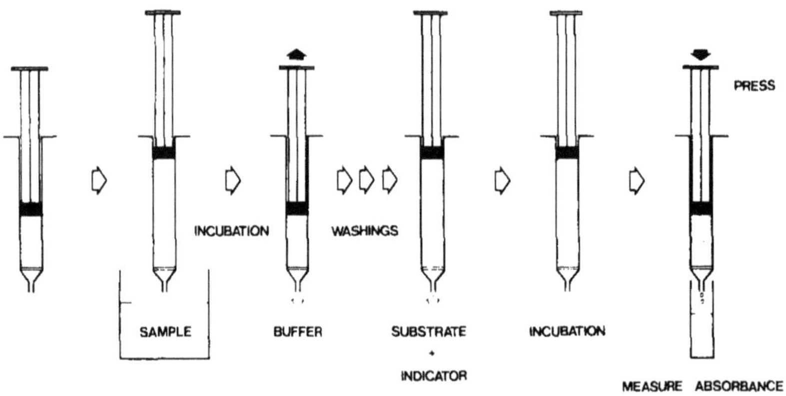

FIGURE 6a. Schematic presentation of an assay using disposable plastic syringes with specific sorbents. The sample is sucked up in the syringe and the cells present are allowed to bind to the immobilized biomolecules (lectins or antibodies) during the first incubation step. The free solution is then pressed out of the syringe and the adsorbent is repeatedly washed by filling the syringe with buffer and emptying it. A substrate solution containing an appropriate indicator is afterwards added and incubated. After this second incubation the solution in the syringe is analyzed by measuring the absorbance change caused by the indicator as a result of the metabolism in the syringe during the second incubation step. (From Mattiasson, B. and Johansson, P. A., *J. Immunol. Methods*, in press. With permission.)

B. Operational Conditions

As seen in many examples in this chapter, the lack of specificity is a severe problem when using immobilized cells in analysis. Thus, in serious efforts to develop these assays much interest has been focused on this problem. The choice of transducer will reduce the effect of potentially disturbing substances. Thus, with the thermal registration, *all* heat generated or consumed by *all* reactions will be registered. This means

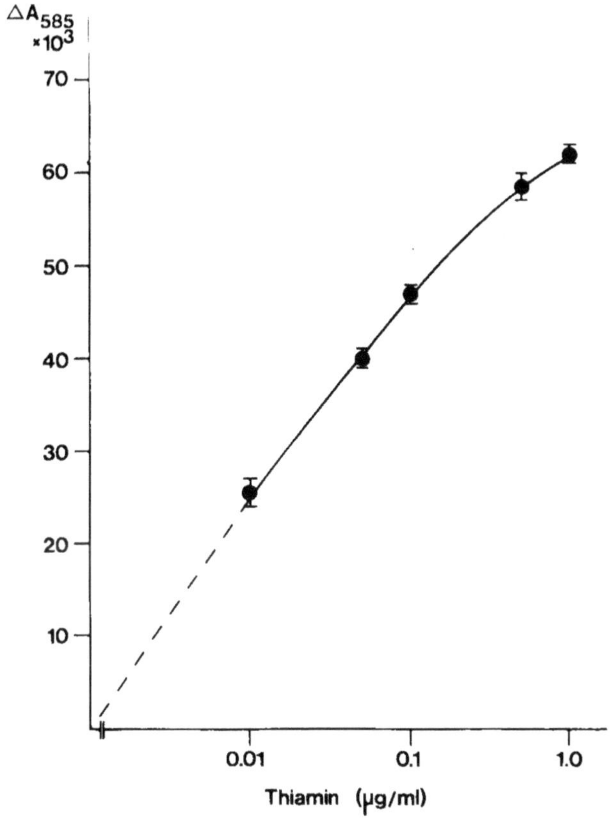

FIGURE 6b. Calibration curve for thiamine. The metabolic response registered as a function of thiamine concentration. All measurements were performed in triplicate. The bars in the figure denote the range of variation of the results of the assays. The syringes used were all preexposed to 6 million *Saccharomyces cerevisiae* cells. The figures are corrected for blank reactions registered in the blank sample. (From Mattiasson, B. and Durholt, M., in press. With permission.)

that in overall assays of metabolic activity such a transducer is directly a good choice,[10] whereas when specific assays are to be performed, certain specific precautions must be taken in every individual case. pO_2-electrodes are generally not selective enough and will register the metabolic events as soon as anything metabolizable appears. Also here the system must be modified before specific assays are possible.

C. Methods for Introducing Specificity

Membranes permeable to gas but not to solutions can be used to eliminate interferences by molecules other than those to be assayed. This is clearly shown by the so-called acetic acid electrode[46] and the ethanol electrode[47,48] being actually the same electrode with the same microorganism. When the electrode is used at pH 3.5, i.e., below the pK for acetic acid, gaseous acetic acid penetrates the membrane together with ethanol. At pH 3.5, there is thus a poor selectivity. However, when the pH is raised to neutrality acetic acid is above its pK and can then not penetrate the membrane. At this higher pH the electrode is therefore more specific for ethanol.

The conditions under which an assay is performed may be of utmost importance for the specificity of an assay. In an assay for glutamine[49,50] *Escherichia coli* containing glutamate decarboxylase was used and a pCO_2-electrode as a transducer. Under aero-

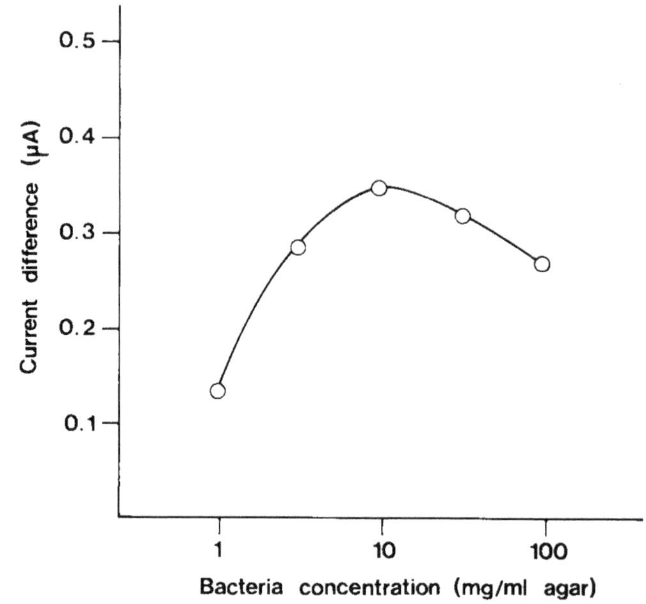

FIGURE 7. Effect of the concentration of *Leuconostoc mesentesoides* entrapped in 2% (w/v) agar on the response to phenylalanine (10^{-5} g/mℓ). The currents were measured after incubation for 90 min at 30°C. (From Matsunaga, T., et al., *Anal. Chim. Acta*, 127, 245, 1981. With permission.)

bic conditions such a combination would be capable of metabolizing a wide range of substances and thus cause a change in pCO_2. However, when operated anaerobically, a much higher selectivity was obtained. More specific electrodes, like ammonia electrodes, are easier to operate. In many of their studies Rechnitz and co-workers[34,51,37] have studied microbe and tissue electrodes for quantification of amino acids. All reports show that at least one or two amino acids will cause several disturbances in every assay. These interferences might not be of practical importance when used under certain conditions and then such electrodes are useful. Also enzyme electrodes have the same effect on several amino acids.[3] However, when the enzyme preparations were properly prepared, i.e., purified with all the techniques available today, pure monospecific enzyme preparations were obtained.[52,53] This suggests that disturbances by other amino acids is not dependent upon catalytic activity in the specific enzyme as such, but merely that the enzyme preparation is contaminated with other amino acid degrading enzymes. To achieve a similar high specificity of cell electrodes as that obtained with highly purified enzymes seems to require selection of mutants with the desired properties.

In many of the cell-based systems diffusion is limiting. Besides the Nernst diffusion layer and the pore diffusion that are ascribed to the immobilization technique, passage through the membrane and metabolic sluggishness may delay the response still more (see Chapter 2, Volume II). Compared with enzyme-based analyses the response time and the regeneration time are longer in cell preparations. To minimize this difference many researchers have used kinetic measurements instead of measurements on the steady state. Even then the cell electrodes are slower, but often more stable.

D. Cell Properties

The metabolic potential of a cell may vary with a wide variety of factors. This makes the cell-electrodes very dependent upon standardized cell preparations, if the results

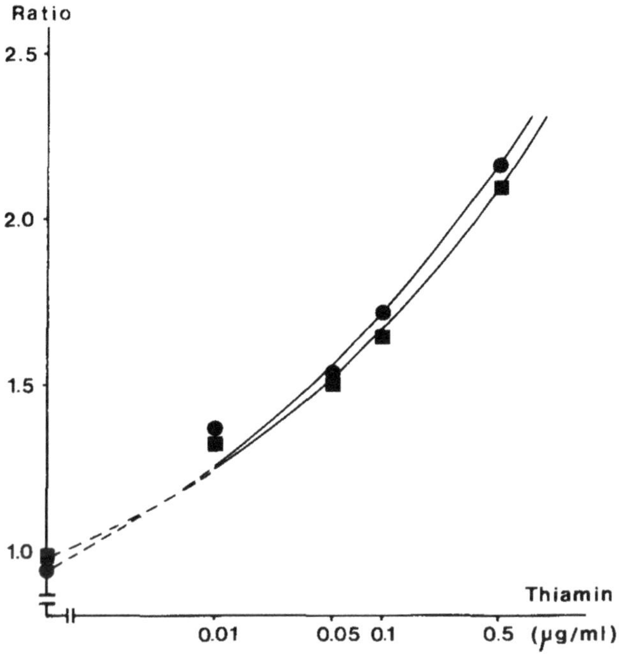

FIGURE 8. Response from yeast-alginate membrane vs. thiamine concentrations. The two curves refer to two sets of measurements made one week apart. The data points on y-axis refer to the value for thiamine concentration = 0. Ratio = ratio of oxygen electrode response after and before exposure of the membrane to thiamine. (From Mattiasson, B., et al., *Enzyme Microb. Technol.*, 4, 153, 1982. With permission.)

are to be reproducible. In bioassays involving the whole cell metabolism or a large part of it, the result is read as the activity after exposure over activity after a blank exposure. This way of expressing the analytical result as a ratio requires that the basal metabolism to be at a constant level before the assay is started. Since each assay involves new cells, no calibration between the assays is possible which makes the constancy of the level more important.

In assays of activators or inhibitors the cells must be depleted of the substance in question. This was shown[12] when assaying thiamine by its positive influence on the metabolism of *Saccharomyces cerevisiae* (Figure 8).

1. Adaptation

Preculturing for long time on a limited medium may change the metabolic pattern in such a way that the enzyme of interest in the assay increases in concentration and those of other substances within the cell decreases. When it does, it improves the situation when the assays are performed. Hikuma et al.[50,54] cultivated *Nitrosomonas* for several weeks on a limited medium before cell electrodes for quantification of ammonia were available. Glucose and amino acids showed barely any interference.

2. Enzyme Induction

In assays for phenols Kjellén and Neujahr used *Trichosporon cutaneum*.[55] This yeast was cultivated with phenol present in the medium since the phenol hydroxylase is an induced enzyme. However, in this case enzymes for normal metabolism were still active and present in normal amounts and thus the assays for phenol using the cells in combination with an oxygen electrode were severely disturbed when assimilable sugars were

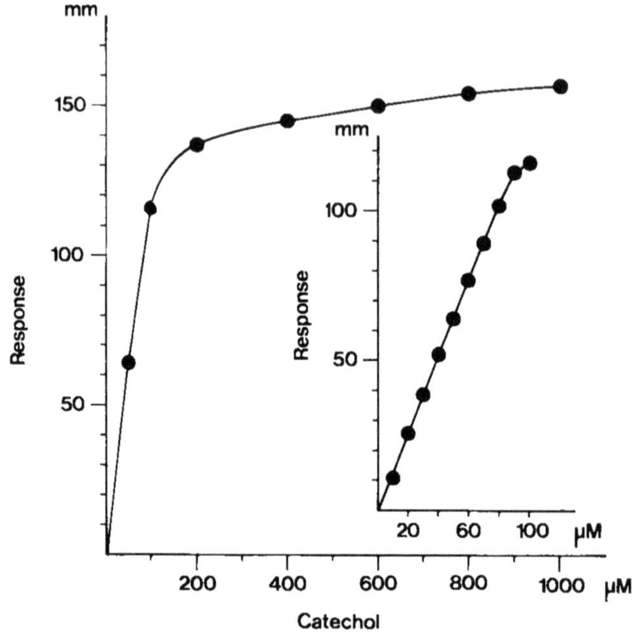

FIGURE 9. Calibration curve for catechol using affinity immobilized *Trichosporon cutaneum*. A small amount of cells were suspended in the perfusion buffer and injected into the flow passing the gel bed in an enzyme thermistor loaded with Concanavalin A-Sepharose. The substrate was then administered as 1 min pulses.

present. Similar results were obtained when the cells were biospecifically reversibly bound to Con A-Sepharose mounted in an enzyme thermistor (Mattiasson, Ramstorp and Neujahr, unpublished) (Figure 9).

Cell growth may be inhibited in the enzyme electrode by the presence of chloramphenicol in buffer and assay medium and this may lead to a drastic decrease in the number of interfering substances, since many enzymes are no longer operating at full capacity.

3. Dead Cells

Another approach may be to kill the cells by destroying the membranes by ultrasound or by lyophilic drying[33,56] or by treatment with acetone.[50] In this study when glutamic acid was analyzed the unspecific influence by glutamine was reduced by 90% with acetone treatment.

4. Enzyme Inhibitors

As briefly touched upon earlier, the background metabolism may cause disturbances. When *Pseudomonas* sp. were used for determination of L-histidine[57] background transamination caused disturbances which were, however, eliminated on addition of an inhibitor of transaminases, isonicotinic acid hydrazide.

V. APPLICATIONS

Immobilized cells can be used for many different analytical purposes. The demands placed on the cells vary with the type of application. The types of specific analyses have been grouped as follows:

- Those involving a single enzyme step
- Those involving coenzymes
- Those involving a large part of the metabolism

Furthermore, some examples of group specific analyses are discussed as well as analyses with no specificity, e.g., BOD-determinations.

A. Single Step Analyses

The use of single enzymes within the cells for a specific analysis restricts the choice of enzymes to those generating or consuming a metabolite detectable with the transducer. Some of the published examples are given in Table 5.

To widen the choice of reactions studied it is also possible to include pre- or post-treatment of the substance to be analyzed with an additional enzyme co-immobilized with the cell (see Chapter 5, Volume II).

B. Coenzyme-Dependent Reactions

Of the analyses based on one of a few enzymic steps in the metabolism of the cell a few require coenzymes (Table 6). Such analyses are very difficult if carried out in free solution[58] and then the use of whole cells offers certain advantages.[59] A serious problem when performing coenzyme-dependent reactions with immobilized enzyme preparations is the recycling and retainment of coenzymes. This problem has not yet been satisfactorily solved and in some cases free coenzyme was therefore used in each assay. Furthermore, the regeneration of the coenzyme has been the subject of much research, in which immobilized redox-molecules as well as electrochemical reactions on electrode surfaces have been tried.[60,61] So far, no system works satisfactorily for use in routine analysis. One can either use cells capable of originating the coenzyme or add artificial electron acceptors.[33]

C. BOD-Sensors

The biological oxygen requirement is very often used for quantifying organic matter in waste water. In conventional procedures, such an assay takes 5 to 7 days and a quicker method would substantially improve the possibilities of an effective control and also for the chances of preventing excessively high values. Several approaches have been tried to develop BOD-sensors.[10,77-80]

The fact alone that immobilized cell preparations are dense cell populations makes it reasonable to expect fast analyses. In an assay procedure for nicotinic acid[45] immobilized *Lactobacillus arabinosus* was placed in the sample for 1 hr before the potential was read (Figure 10). Thus, in this case no biosensor combination was used, and it seems reasonable to assume that such long-lasting incubations will take place separated from the transducer and that registration will be performed in a subsequent step when the transducer is momentarily in contact with the incubation solution.

The same situation was used with *Saccharomyces cerevisiae* reversibly bound to concanavalin A-Sepharose and deposited in a disposable plastic syringe.[40] After a preset period of time, the incubation solution was pressed out of the sorption material and collected in a cuvette and read in a spectrophotometer. Thiamine and amphoteracin were quantified this way (Figure 11).[41]

Assays based on metabolism of the whole cell may also be used for quantifying toxic substances. It is then possible under well-defined conditions to quantify a certain toxic substance or to quantify the toxic effect of mixtures of toxic substances whose exact composition is not known and whose toxicity may vary widely. Using this approach a toxiguard was constructed to be placed in connection with the biobed in plants for the treatment of waste water.[24] The toxiguard consisted of an oxygenator, a biostep con-

Table 5
ANALYTICAL APPLICATIONS OF IMMOBILIZED CELLS UTILIZING ONE OR A FEW ENZYME CATALYZED STEPS

Metabolite	Organism	Transducer	Interference	Sensitivity	Comments	Ref.
Aspartate	*Bacterium cadaveris* ATCC 9760	NH_3-gas sensing electrode	Adenosine asparagine	10^{-4} — 10^{-2}	Stable 10 d, regenerable	34
Glutamine	Porcine kidney cortex		Not reported	-3×10^{-5}	Stable 30 d	26
	Sarcina flava ATCC 147		Asp, his	10^{-4} — 10^{-2}	Stable 2 w.	35
	Porcine kidney cortex			-7×10^{-5}	> 28 d	36
Cystein	*Proteus morganii*	H_2S-electrode	Homocystein cystathionine (electrode does not discriminate H_2S from CO_2)	5×10^{-5} — 9×10^{-4}	6 d	62
Serine	*Clostridium acidiurici*	NH_3-gas sensor	L-glutamine	$1.8\ 10^{-4}$ — $1.6\ 10^{-2}$	3 d	63
Histidine	*Pseudomonas* sp.	NH_3-gas sensor	Urocanic acid, urea	10^{-4} — 3×10^{-3}	3 w	57
Arginine	*Streptococcus faecium* ATCC 9790	NH_3-gas sensor	Glutamine, aspargine	5×10^{-5} — 10^{-3}	20 d	51
Phenylalanine	*Leuconostoc mesenteroides*	pO_2-electrode	N.R.	10^{-7} — 5×10^{-5}	Cells + lactate oxidase electrode	64
Phenol	*Tricosporon cutaneum*	pO_2-electrode	Other phenols	1 — 20 mg/ℓ	Stable > 5d regenerable	59
NAD^+	*Escherichia coli*	NH_3-gas electrode	Nicotinamide	2.5×10^{-4} — 2.5×10^{-3}	Hybride sensor (enzyme + bacteria)	65
Cholesterol	*Nocardia erythropolis*, IMET 7185	pO_2-electrode	Glucose	15—130 μM	Stable > 4w response time 35—70 sec	66
3-keto-4-ene steroid	*Nocardia opaca* IMET 7030	Rotating glassy carbon electrode	N.R.	2.5 — 100 μM	Artif. electr. acceptor	67
L-lactate	*Hanensula anomala*	pO_2-electrode	Glucose DL-lactate pyruvate sucrose ethanol		Intact cells, artif. el. accept.	33
L-Lactate			—		Lyzed cells, artif. el. accept.	56
Cephalosporins	*Citrobacter freundii*	pH-electrode	N.R.	50—300 μg/mℓ 7-ADAC	Stable 1 w.	68

Table 6
BIOSENSORS BASED ON COENZYME SINGLE OR MULTISTEP REACTIONS

Metabolite analyzed	Ref.
Phenol	59
Nitrate	69
L-lactate	33

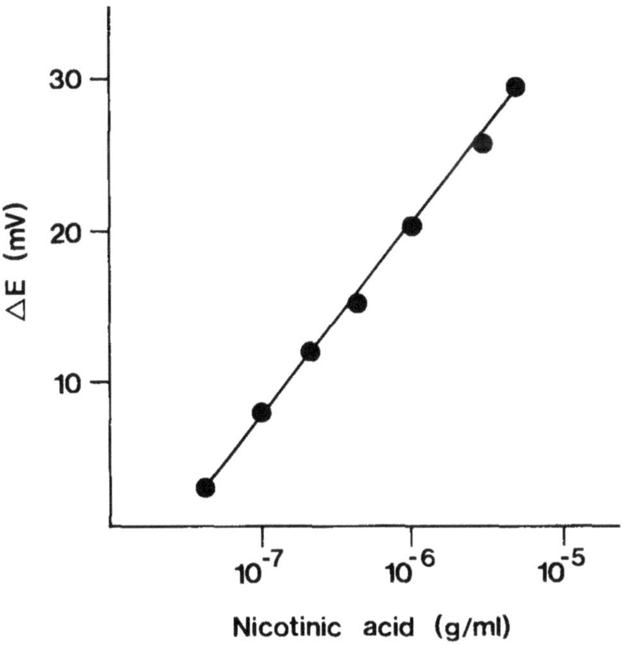

FIGURE 10. Calibration curve for nicotinic acid when incubating the sample to be analyzed for 1 hr with high amounts of *Lactobacillus arabinosus* before the potential change was read. (From Matsunaga, T., et al., *Anal. Chim. Acta,* 99, 233, 1978. With permission.)

taining a submerged biological filter with rotational plates covered with a thin film of living microorganisms and in the effluent from the filter an pO_2-sensor. When operated in conjunction with a biobed, the toxiguard contains the same mixed culture of microorganisms as the biobed. It has proved possible to detect heavy metal ions as well as cyanide with such a detector. If the poisoning is not too severe, the culture within the biobed recovers within 30 min to a few hours after the poison has been washed out (Figure 12).

Uncoupling substances can be quantified. Since ATP-production is stopped in the terminal electron transport chain less energy is conserved as metabolic energy and more heat is generated instead. This was shown with the use of a microbe thermistor treated with glucose containing buffer pulsewise supplied with 2,4-dinitrobenzene. It is difficult to find an equivalent to this generation of heat when using other general transducers for quantification (Figure 13).

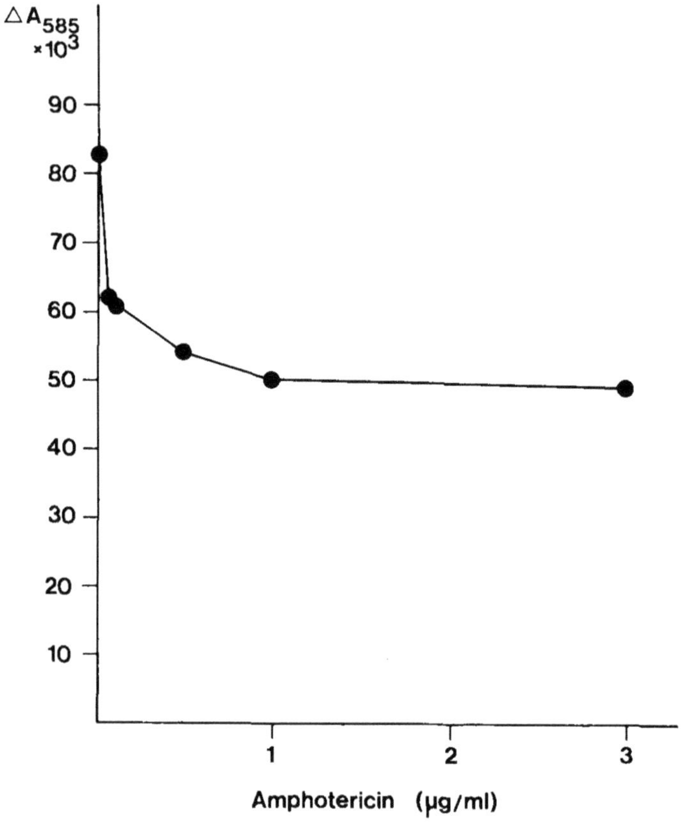

FIGURE 11. Effect of exposure to varying concentrations of amphotericin on the metabolic response from cells of *Saccharomyces cerevisiae* affinity bound to Concanavalin A in the disposable plastic syringe assay. (From Mattiasson, B. and Dürholt, M., in press. With permission.)

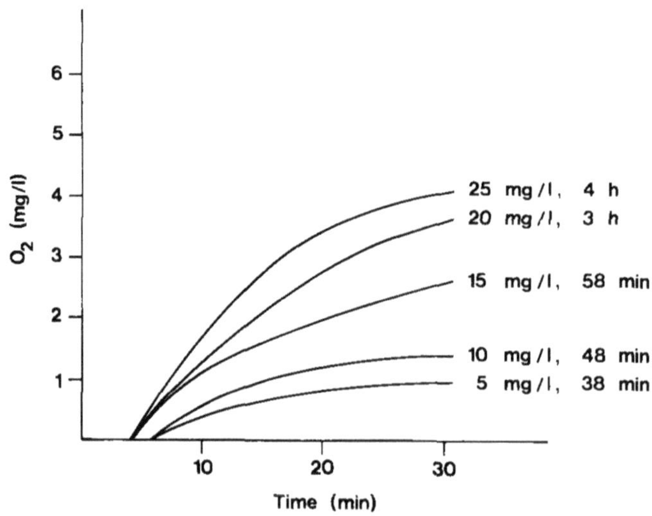

FIGURE 12. Toxicity indication for cyanide from a continuous monitoring system upon administration of pulses of cyanide to the system. The time periods given indicate the time needed for the system to recover. (From Solyom, P., *Vatten*, 2/76, 192, 1976. With permission.)

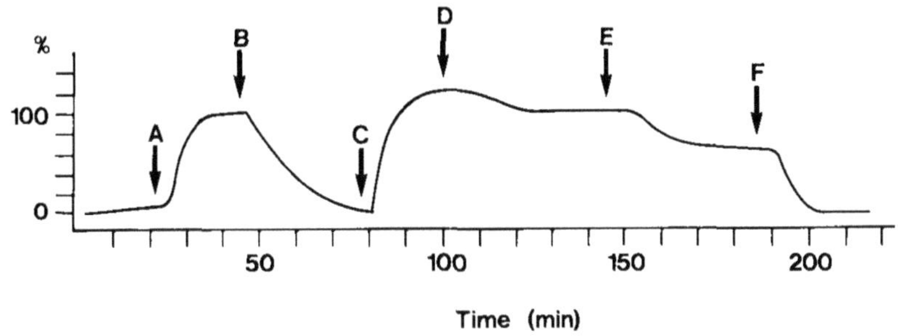

FIGURE 13. Thermogram obtained with the microbe thermistor, containing yeast cells entrapped in polyacrylamide following introduction of various metabolites and inhibitors dissolved in 0.1 *M* potassium phosphate buffer pH 7.0 into the flow. The arrows indicate addition of (A) 1 m*M* glucose; (B) buffer; (C) 1 m*M* glucose + 1 m*M* 3,4-dinitrophenol; (D) 1 m*M* glucose; (E) 1 m*M* glucose + 2 m*M* arsenate; (F) buffer. The steady-state response to 1 m*M* glucose is set as 100%. (From Mattiasson, B., et al., *Nature (London)*, 268, 519, 1977. With permission.)

D. Analyses of Hormones and Other Specific Effectors of Cell Metabolism

Before the development of immunochemical binding analyses much time and money was spent on analyzing hormones, e.g., with conventional bioassays. Such assays have now been completely ousted by immunochemical analyses.

The introduction of simple and effective methods for culturing cells seems to appear new possibilities for bioassays. It is thus possible to use cultured cells instead of whole animals in such assays. So far, only very little work has been reported in this area, but the potential of combining cell lines and biosensor technology looks very promising. Analysis of insulin by immobilized lymphocytes with the use of an oxygen sensor is still in a very early stage of development.[25] The possibility of quantifying receptors on the cell surface also seem obvious.

E. Group Specific Analyses

Most researchers have found the low specificity shown by some cells for the substrate as a serious obstacle. However, in some cases it may be advantageous to measure the whole group of substances instead of each one separately. The basic requirement for such an assay is that either a single enzyme has a broad specificity, or several different enzymes generate common products that flow through the main metabolic pathways and thus generate signals detectable with a general transducer.

Another alternative is that the group of substances to be quantified have a similar influence on the metabolism without being converted. For otherwise a secondary effect due to metabolism of another substrate may be registered. Such a useful secondary effect can be caused by the deleterious action of detergents on biological membranes[28] or of various toxic substances on the ability of neuroblastoma cells to attach to a solid surface.[72] In these last two examples the method does not quantify the substances as such but merely their biological effects on the specific target used.

Table 7 gives some of the systems characterized by such a broad substrate tolerance that they may at least be used for estimating the content of members of a group of chemicals. To improve this broad substrate tolerance still more, mixed cultures (Chapter 4, Volume I) or mixtures of enzymes and microbes can be used (Figure 14).

F. Analysis Based on a Major Part of the Cell Metabolism

Bioassays of vitamins and other effector substances are based on the whole metabolism of the cells. In the conventional performance, cell divisions are necessary before the result can be read, and when using biosensors the metabolic prelude to cell division

Table 7

Sensor	Group of substances	Ref.
Sugar electrode	Assimilable sugars	73
Plaque electrode		74
Gluconobacter electrode		75
Cell electrode	Assimilable substance	76
Enzyme electrode in combination with immobilized liposomes	Surface active substances	28,29
Particle counter and immobilized neuroblastoma cells	Organic solvents, heavy metals	72

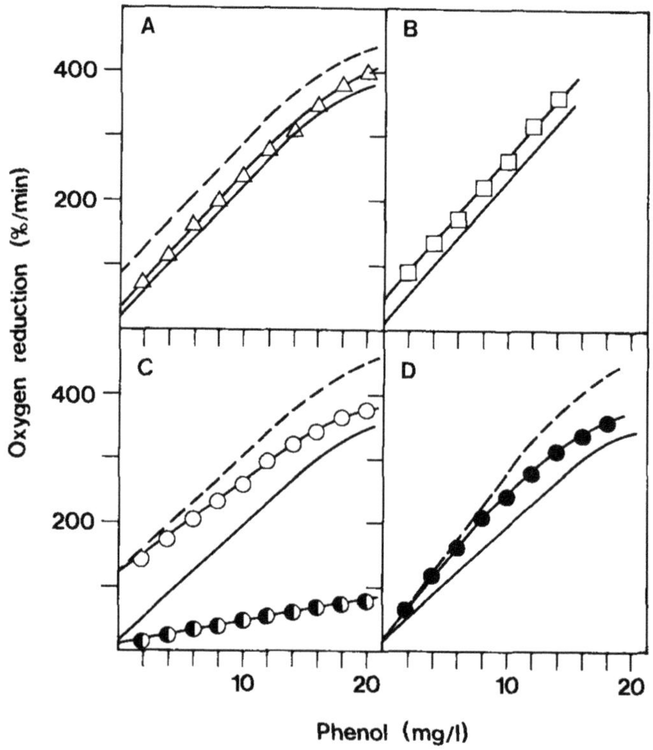

FIGURE 14. Phenol response of bioprobe (Clark oxygen electrode, coated with cell paste of phenol-induced *Trichosporon cutaneum* (—) Phenol alone. (— — —) Calculated curve if response to phenol is added to that of other sample components. (A) (△) Phenol in presence of 200 mg glucose/*l*. (B) (□) Phenol in presence of 456 mg/*l* (anhydrous) sodium acetate. (C) Mixture of eight phenol derivatives including cresols, chlorophenols, resorcinol, and catechol in equimolar amounts (◐); phenol in the presence of 36 mg/*l* of the same mixture (○). (D) Mixture of the eight phenol derivatives together with an equimolar amount of phenol (●). (From Neujahr, H. Y. and Kjellen, K. G., *Biotechnol. Bioeng.*, 21, 671, 1979. With permission.)

is registered and used to deduce the concentration in the sample.[70] Such assays have been reported only for a small number of substances. All the cells used so far have been microorganisms, but it seems realistic to expect that with time also mammalian and plant cells will be exploited in specific assays. As deduced from conventional bioassays, this technique is potentially very sensitive. Some calculations point towards as-

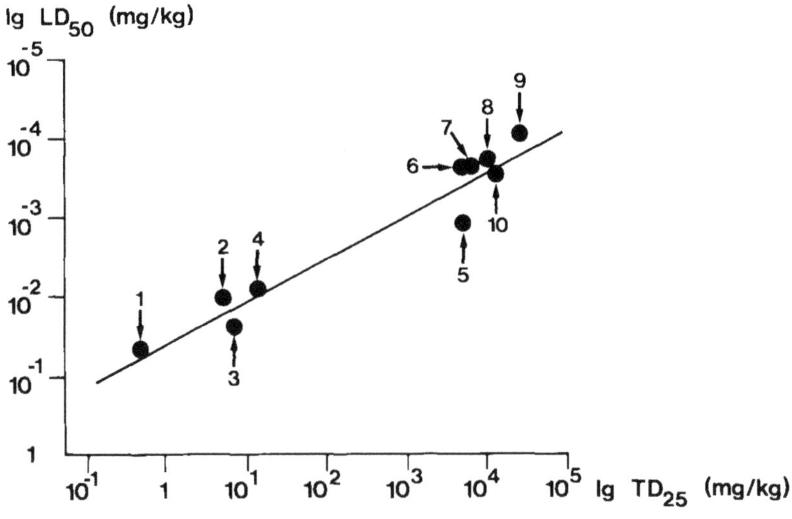

FIGURE 15. *Correlation between LD_{50} values (rat, oral administration, values from Registry of Toxic Effects of Chemical Substances 1979 and "Metals" National Swedish Environment Protection Board, 1976) and TD_{25} values for CH_3HgCl (1), $CdCl_2$ (2), $HgCl_2$ (3), ppDDT (4), n-butanol (5), benzene (6), 1982) dioxan (7), n-propanol (8), acetone (9) and t-butanol (10). The logarithm of the LD_{50} values were plotted vs. the ng logarithm of the TD_{25} values and the least-squares fit of a straight line through the points was calculated. Equation of the line: $y = 0.54 \cdot x + 1.41$. Correlation coefficient: 0.98. m. (From Peterson, A., et al., *Toxicol. Lett.*, 9, 101, 1981. With permission.)*

says as sensitive as RIA.[71] However, in other biosensor assays the sensitivity has been increased at the expense of the quickness of the analyses. A sensitive assay takes longer than a less sensitive one. The sensor-based assays can be used for relatively well-defined waste waters. The response time is in the order of 20 min.

G. Analyses Based on Changes in the Physical Properties of Cell Membranes

Most of the work discussed so far has been carried out on microbial cells, but a few efforts to use mammalian cells have also been made. Since many mammalian cells grow on a solid support it is quite easy to obtain an immobilized cell preparation. On the other hand, cells grown on a surface are exposed to the environment much more than are entrapped cells. One application that has become important is the use of cells growing on the bottom of the wells in microtiter plates for studying the effect of virus infection. When virus infection takes place cell lysis occurs, and this can be observed rather easily. By using dilution series of the samples to be analyzed a certain titer of the virus content can be estimated, i.e., the highest dilution in which lysis will still appear in the immobilized cell preparation.[81] The ability of at least some cell lines to adhere to the surface requires favorable conditions. When the cells lose activity and die, they desorb from their support. This effect may be used as a measure of the toxicity of various compounds added to the medium in which the immobilized cells appear. Walum and Peterson[82] and Peterson et al.[72] used neuroblastoma cells growing on Falcon petri dishes. After exposure of the cells to the samples to be analyzed the number of suspended and attached cells were counted in a Coulter Counter. In such an assay it is assumed that the number of cells used is constant. When working with many mammalian cells this is not difficult since they only grow as monolayers on the surface available (Chapter 4, Volume I). Spontaneous cell desorption causes disturbances, but the method is nevertheless good for screening the conditions of a cell preparation exposed to different chemicals. A value comparable to the LD_{50} value for animals has been established (Figure 15).

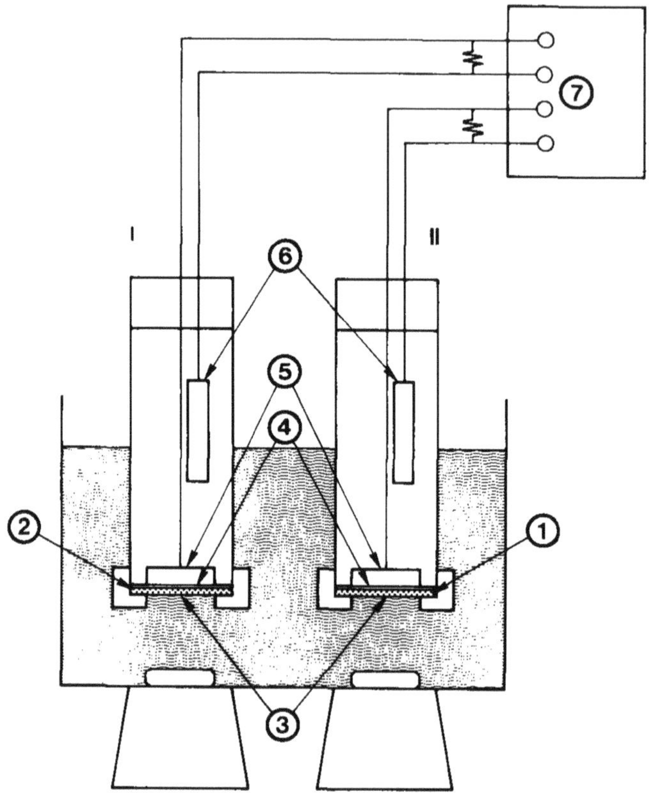

FIGURE 16. Schematic diagram of the electrode system for rapid detection of chemical mutagen: (I) Rec⁻ electrode, (II) Rec⁺, (1) *Bacillus subtilis* Rec⁺, (2) *Bacillus subtilis* Rec⁻ (3) membrane filter, (4) teflon membrane, (5) Pt cathode, (6) Pb anode, (7) recorder. (From Karube, I., et al., *Anal. Chem.*, 53, 1024, 1981. With permission.)

Another feature of the cells that die is a change in the membrane properties. This is routinely used in hematology when staining with trypan blue for assessing the proportion of dead cells in a preparation.[83] This change of uptake can be utilized in an assay. By using a monolayer of living cells and exposing it to certain chemicals prior to addition of the dye it is possible to estimate the extent of damage to the membranes.

Using microorganisms Mattiasson and Dürholt[41] succeeded in demonstrating the same phenomenon for immobilized yeast cells. Treatment of the cells with amphoteracin prior to addition of a glucose solution containing the redox-indicator methylene blue showed that treated cells accumulated more redox indicator than did less treated cells. The phenomenon was used in a quantitative assay of amphoteracin, but can be used for assaying a wide variety of substances interfering with cell membranes (Figure 11). A similar approach was also taken for rapid determination of the viability of yeast cells in free solution.[84]

H. Screening of Mutagens

The high density of cells in an immobilized preparation can be utilized when screening for mutagens. Karube et al.[85] used two strains of *Bacillus subtilis* immobilized in separate porous acetyl cellulose membranes and mounted on separate oxygen sensors. One strain was of a wild type and the other was deficient in the DNA recombination enzyme system (Figures 16 and 17).

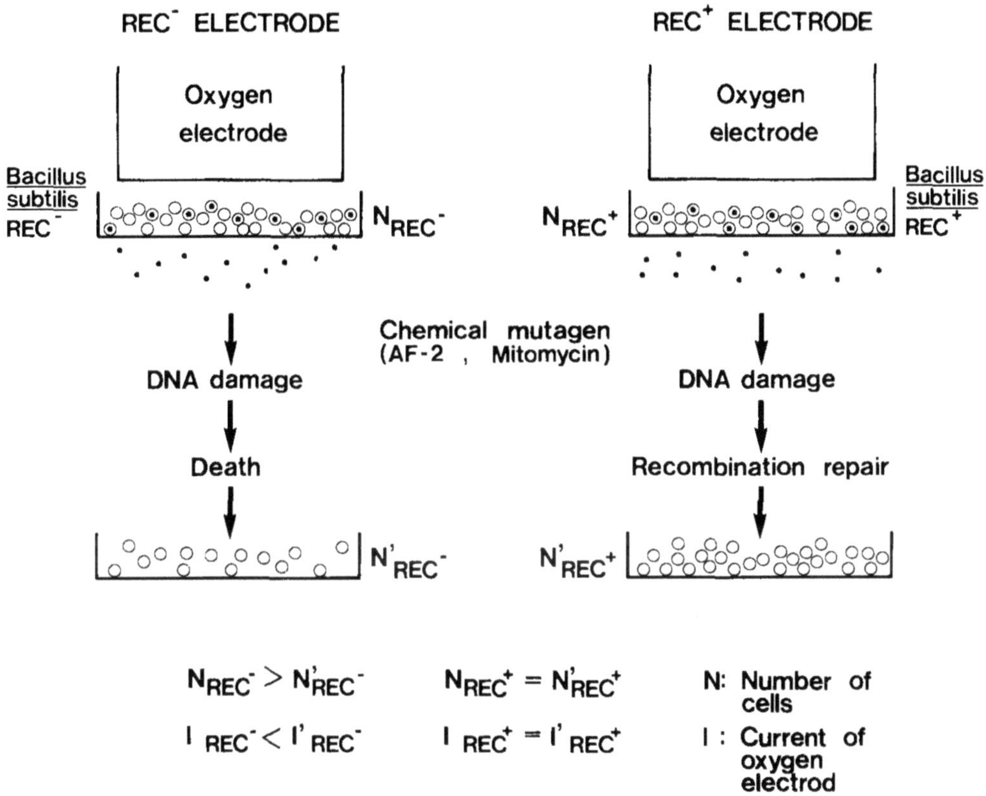

FIGURE 17. Principle of the microbial electrodes for preliminary screening of mutagens. (From Karube, I., et al., *Anal. Chem.*, 53, 1024, 1981. With permission.)

The current between the two electrodes was measured and changes were observed when the cells were exposed to mutagens. The changes observed were ascribed to the effect of mutagens on the deficient strain, because these cells died from lack of repair system after mutation, whereas the wild type remained stable. The assay system was tested with several well-known mutagens and a response occurred within 1 hr. However, when inhibitors of cell metabolism or protein synthesis were added, no differences were observed. The assay system proved more sensitive to the mutagens tested than Ames test and others used today. Furthermore, the response time was much shorter.

I. Cell Counting

No quick and simple methods are available for counting cells. The method most often used is plate counting, which is laborious and tedious and takes a long time, often 2 days or more. In an effort to speed up this assay Sakato et al.[86] developed an electrode which was directly sensitive to the number of cells in the solution. The mechanism at work in this system is not fully understood. A signal proportional to the number of cells was found to be in the range 10^8 to 4×10^9 for bacterial cells/mℓ and 10^3 — 10^6 cells/mℓ for mammalian cells. This method which is not directly based on immobilized cells is not at all selective. The signal denotes the number of all cells present in the sample. As in fermentations using monocultures, this limitation is of no practical importance.

Another approach was taken by Mattiasson and Johansson,[40] who used biospecific sorbents to reversibly immobilize cells before exposure to substrate and a suitable indicator. When operated under standardized conditions the method was very sensitive

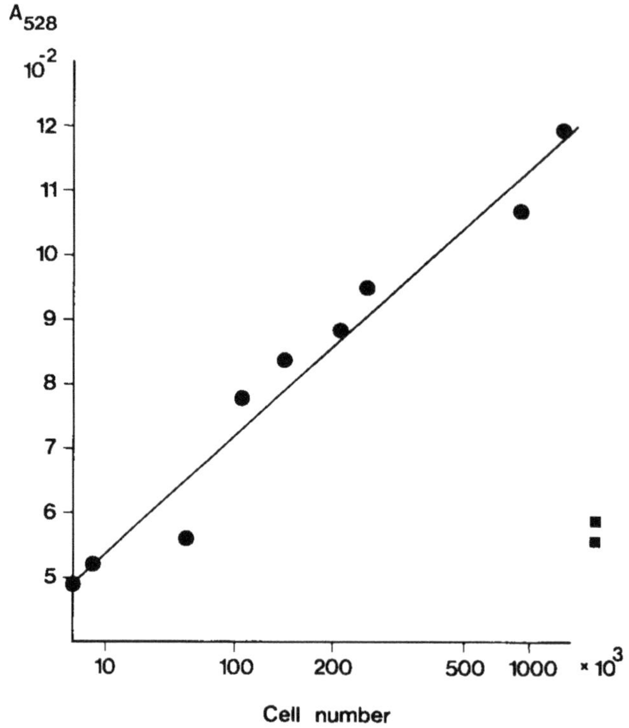

FIGURE 18. Calibration curve for quantification of yeast cells according to the principle demonstrated in Figure 6a. (From Mattiasson, B. and Johansson, P. A., *J. Immunol. Methods*, 52, 233, 1982. With permission.)

and selective. In studies of yeast cells *Saccharomyces cerevisiae* 50,000 cells were counted. In other assays specific sorbents with antibodies against certain bacteria were used and the sensitivity of the method was also high. Besides this disposable syringe, a similar device for flow systems was developed. The fact that it is the number of cells exposed to the sorbent that is important and not the concentration makes it possible to collect cells during a relatively long period before assay and in such a way obtain a very accurate measure of the cell concentration. During long periods of exposure it is very important that already bound cells adhere to the sorbent and do not desorb. When working with cells the risk of desorption is much less than when working with single macromolecules.[87] This is because the cells are bound by multipoint attachment, which means that even if the interaction *per se* is not so strong, the cells will still be firmly bound.[15]

The use of pH-indicators to assess the number of cells is based on the phenomenon that cells can normally convert neutral compounds, such as glucose, to acid products (Figure 18).[40] This same phenomenon was exploited by Fleishaker et al.[38] when developing a conductivity-based analytic method for growth of human fibroplasts. A linear correlation between conductance and cell number was found between 0.2 — 1.3 10^6 cells/mℓ.

J. Affinity Immobilization of Cells

Affinity purification techniques for soluble biomolecules are available.[88] Recent application of the same principles to cell purification has given positive resolution of cell mixtures.[89] The quantification of the different species may be performed by a counter-

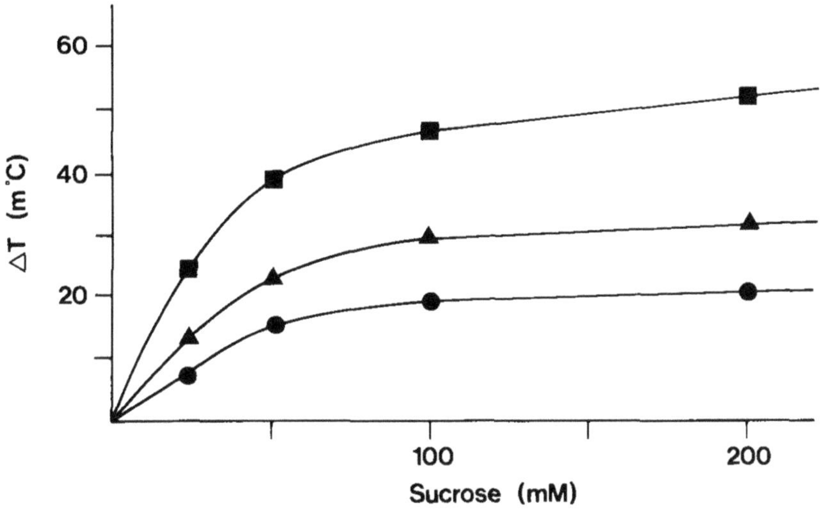

FIGURE 19. Heat response in a microbe thermistor with alginate entrapped yeast; (■) without ethanol present; (▲) with 5% ethanol present and (●) with 11% ethanol. (From Mandenius, C. F., et al., *Biotechnol. Lett.*, 3, 629, 1981. With permission.)

current technique. Another alternative is to supply these affinity immobilized cells with substrate and then to quantify them from their metabolic activity.[90] The resolution in such simple systems is high.

K. Immobilized Cells in Fermentation Control

A crucial point in fermentation technology is the monitoring of the events in the fermenter. The usual procedure used is a registration of parameters such as pH, pO_2, pCO_2, viscosity, and turbidity. Based on these data a picture of the real situation in the fermenter is deduced. With the introduction of biosensors it has been possible also to follow changes in the concentration of essential metabolites, e.g., substrates and products.[50,90-92] Thus, enzyme electrodes or enzyme thermistors can monitor penicillin fermentation more precisely than formerly. In systems where product inhibition takes place, a crucial point besides concentration of substrate and product is also the conditions under which the cells operate, i.e., the severity of the inhibition. To elucidate this question the fermentation of sucrose to ethanol by immobilized *Saccharomyces cerevisiae* was studied, where the sucrose concentration was controlled by an enzyme thermistor unit. By using a biosensor based on an immobilized preparation of the same species as is kept in the fermenter, it was possible to follow the buildup of the product in the medium and the concomitant inhibition of the activity of the yeast cells (Figure 19).[82]

This approach is similar to that applied in the toxiguard discussed earlier. With this kind of device it may be possible to monitor inhibiting concentrations of products as well as any other factor hampering the catalytic effectiveness of the immobilized cells.

VI. CONCLUDING REMARKS

In the last few years the exploitation of immobilized cells for analytical purposes has been rapid and the range of its applications is widening. The problem of specificity is still not solved. One can predict that even if the cell-based analytical systems have many advantages over enzyme-based electrodes, there will still be a good market for enzyme-based sensors because of their higher specificity. On the other hand, in some cases it is very tempting to exploit the greater stability of enzymes *in situ* when stability of the free enzymes is extremely poor. In such cases it must be possible to operate with the whole range of conditions of the cells from actively dividing cells — resting cells — permeabilized cells — dead cells.

Furthermore, cells may play an important role in the measurement of effects on biological systems. Such applications as environmental analysis for surface substances, mutagenic substances, toxic substances, BOD, etc., seem to be rather probable practical applications; in some specific cases units are already operating. The application of immobilized cells in analysis is still in its infancy and a broad field of applications can be foreseen in the near future.

REFERENCES

1. **Bergmeyer, H. U.,** *Methoden der Enzymatischen Analyse,* Verlag Chemie, 1969.
2. **Guilbault, G. G.,** *Handbook of Enzymatic Methods of Analysis,* Marcel Dekker, New York, 1977.
3. **Guilbault, G. G.,** Immobilized enzymes as analytical reagents, *Appl. Biochem. Biotechnol.,* 7, 85, 1982.
4. **Clark, L. C., Jr.,** A family of polarographic enzyme electrodes and the measurement of alcohol, *Biotechnol. Bioeng. Symp.,* 3, 377, 1972.
5. **Maggio, E. T., Ed.,** *Enzyme Immunoassay,* CRC Press, Boca Raton, Fla., 1980.
6. **Parker, C. N.,** *Radioimmunoassay of Biologically Active Compounds,* Prentice-Hall, Englewood Cliffs, N.J., 1976.
7. **Chibata, I., Tosa, T., Sato, T., Mori, T., and Yamamoto, K.,** Continuous enzyme reactions by immobilized microbial cells, in *Enzyme Engineering,* Vol. 2, Pye, E. K. and Wingard, L. B., Jr., Eds., Plenum Press, New York, 1974, 303.
8. **Enfors, S.-O. and Nilsson, H.,** Design and response characteristics of an enzyme electrode for measurement of penicillin in fermentation broth, *Enzyme Microb. Technol.,* 1, 260, 1979.
9. **Broun, G. G.,** Chemically aggregated enzymes, *Methods Enzymol.,* 44, 263, 1976.
10. **Mattiasson, B., Larsson, P.-O., and Mosbach, K.,** The microbe thermistor, *Nature (London),* 268, 519, 1977.
11. **Veliky, I. A. and Williams, R. E.,** The production of ethanol by *Saccharomyces cerevisiae* immobilized in polycation-stabilized calcium alginate gels, *Biotechnol. Lett.,* 3, 275, 1981.
12. **Mattiasson, B., Larsson, P.-O., Lindahl, L., and Sahlin, P.,** Vitamin analysis with use of a yeast electrode, *Enzyme Microb. Technol.,* 4, 153, 1982.
13. **Matsunaga, T., Karube, I., Teraoka, N., and Suzuki, S.,** Rapid determination of phenylalanine with immobilized *Leuconostoc mesenteroides* and a lactate electrode, *Anal. Chim. Acta,* 127, 245, 1981.
14. **Mattiasson, B.,** Reversible immobilization of enzyme with special reference to analytical applications, *J. Appl. Biochem.,* 3, 183, 1981.
15. **Mattiasson, B. and Lindahl, L.,** A biosensor for identification and quantification of microbial cells, submitted for publication.
16. **Borrebaeck, C., Börjesson, J., and Mattiasson, B.,** Thermometric enzyme-linked immunosorbent assay in continuous flow system: optimization and evaluation using human serum albumin as a model system, *Clin. Chim. Acta,* 86, 267, 1978.
17. **Mattiasson, B., Svensson, K., Borrebaeck, C., Jonsson, S., and Kronvall, G.,** Non-equilibrium enzyme immunoassay of gentamicin, *Clin. Chem.,* 24, 1770, 1978.
18. **Green, N. M.,** Avidin, I. The use of ^{14}C biotin for kinetic studies and for assay, *Biochem. J.,* 89, 585, 1963.
19. **Lancet, D., Isenman, D., Sjödahl, J., Sjöquist, J., and Pecht, I.,** Interactions between staphylococcal protein A and immunoglobulin domains, *Biochem. Biophys. Res. Commun.,* 85, 608, 1978.
20. **Hubbard, A. L. and Cohn, Z. A.,** *Biochemical Analysis of Membranes,* Maddy, A. H., Ed., Halsted Press, New York, 1976, 427.
21. **Mattiasson, B. and Borrebaeck, C.,** Novel approaches to enzyme-immunoassay, in *Enzyme Immunoassay,* Maggio, E. T., Ed., CRC Press, Boca Raton, Fla., 1980.
22. **Tosa, T., Mori, T., Fuse, N., and Chibata, I.,** Continuous enzyme reactions. I. Screening of carriers for preparation of water-insoluble aminoacylase, *Enzymologia,* 31, 214, 1966.

23. Dahlgren-Caldwell, K., Axen, R., and Porath, J., Utilization of hydrophobic interaction for the formation of an enzyme reactor bed, *Biotechnol. Bioeng.*, 17, 613, 1975.
24. Solyom, P., Kontinuerlig overvakning av akuttoxicitet i avloppsvatten, *Vatten*, 2/76, 192, 1976.
25. Mattiasson, B., Biospecific reversible immobilization — a way to introduce labile structures in analytical systems, *Appl. Biochem. Biotechnol.*, 7, 121, 1982.
26. Meyerhoff, M. E., Preparation and response properties of selective bioelectrodes utilizing polymer membrane electrode-based ammonia-gas sensors, *Anal. Lett.*, 13(B15), 1345, 1980.
27. Canty, T. G. and Wunderlich, J. R., Quantitative *in vitro* assay of cytotoxic cellular immunity, *J. Natl. Cancer Inst.*, 45, 761, 1970.
28. Eriksson, H. and Mattiasson, B., Quantitation of detergents based on their ability to solubilize enzyme-containing liposomes, *J. Appl. Biochem.*, 2, 241, 1980.
29. Eriksson, H. and Mattiasson, B., A liposome based assay for quantification of detergents, *Appl. Biochem. Biotechnol.*, 7, 81, 1982.
30. Mattiasson, B. and Eriksson, H., An assay of detergents based on stabilized liposomes, submitted.
31. Johnston, H. H. and Newsom, S. W. B., Ed., *Rapid Methods and Automation in Microbiology*, Research Studies Press, Inc., Forest Grove, Oregon, 1977.
32. Rytel, M. W., *Rapid Diagnosis in Infectious Disease*, Rytel, M. W., Ed., CRC Press, Boca Raton, Fla., 1979.
33. Kulys, J. J., The development of new analytical systems based on biocatalysts, *Anal. Lett.*, 14(B6), 377, 1981.
34. Kobos, R. U. and Rechnitz, G. A., Regenerable bacterial membrane electrode for L-aspartate, *Anal. Lett.*, 10(10), 751, 1977.
35. Rechnitz, G. A., Riechel, T. L., Kobos, R. K., and Meyerhoff, M. E., Glutamine-selective membrane electrode that uses living bacterial cells, *Science*, 199, 440, 1978.
36. Rechnitz, G. A., Arnold, M. A., and Meyerhoff, M. E., Bioselective membrane electrode using tissue slices, *Nature (London)*, 278, 446, 1979.
37. Arnold, M. A. and Rechnitz, G. A., Comparison of bacterial, mitochondrial, tissue and enzyme biocatalysts for glutamine selective membrane electrodes, *Anal. Chem.*, 52, 1170, 1980.
38. Fleischaker, R. J., Giard, G. J., Weaver, J., and Sinskey, A. J., Progress with computor-coupled mammalian cell culture investigations, in *Adv. in Biotechnol.*, Vol. 1, Moo-Young, M., Robinson, C. W., and Vezina, C., Eds., Pergamon Press, Toronto, 1981, 425.
39. Mattiasson, B., Danielsson, B., and Mosbach, K., Applications of the enzyme thermistor in analysis and process control, in *Food Process Engineering*, Vol. 2, Linko, P. and Larinkari, J., Eds., Allied Science Publishers, London, 1980, 59.
40. Mattiasson, B. and Johansson, P.-A., A simple and quick method for quantifying microorganisms by their metabolic activity when attached to a specific adsorbent, *J. Immunol. Methods*, 52, 233, 1982.
41. Mattiasson, B. and Dürholt, M., A quick and simple bioassay system for effectors of cell metabolism using biospecifically immobilized cells: assay of thiamine and amphotericine using yeast cell, submitted for publication.
42. Weaver, J. C., Reames, F. M., De Alleume, L., Perley, C. R., and Cooney, C. L., Continuous measurements on immobilized cells by a mass filter, *Enzyme Eng.*, 4, 403, 1978.
43. Weaver, J. C., Perley, C. R., Reames, F. M., and Cooney, C. L., Temporarily immobilized microorganisms: rapid measurements using a mass spectrometer, *Biotechnol. Lett.*, 2, 133, 1980.
44. Weaver, J. C., *Biomedical Applications of Immobilized Enzymes and Proteins*, Vol. 2, Chang, T. M. S., Ed., Plenum Press, New York, 1980, 207.
45. Matsunaga, T., Karube, I., and Suzuki, S., Rapid determination of nicotinic acid by immobilized *Lactobacillus arabinosus*, *Anal. Chim. Acta*, 99, 233, 1978.
46. Hikuma, M., Kubo, T., Yasuda, T., Karube, I., and Suzuki, S., Amperometric determination of acetic acid with immobilized *Trichosporon brassicae*, *Anal. Chim. Acta*, 109, 33, 1979.
47. Hikuma, M., Kubo, T., Yasuda, T., Karube, I., and Suzuki, S., Microbial electrode sensor for alcohols, *Biotechnol. Bioeng.*, 21, 1845, 1979.
48. Karube, I. and Suzuki, S., Microbial sensor for ethylalcohol, acetic acid and ammonium, *Enzyme Eng.*, 5, 263, 1980.
49. Hikuma, M., Ooana, H., Yasuda, T., Karube, I., and Suzuki, S., A potentiometric microbial sensor based on immobilized *Escherichia coli* for glutamic acid, *Anal. Chim. Acta*, 116, 61, 1980.
50. Hikuma, M., Yasuda, T., Karube, I., and Suzuki, S., Application of microbial sensors to the fermentation process, *Proc. N.Y. Acad. Sci.*, 369, 307, 1981.
51. Rechnitz, G. A., Kobos, R. K., Riechel, S. J., and Gebauer, C. R., A bio-selective membrane electrode prepared with living bacterial cells, *Anal. Chim. Acta*, 94, 357, 1977.
52. White, W. C. and Guilbault, G. G., Lysin specific enzyme electrode for determination of lysine in grains and foodstuffs, *Anal. Chem.*, 50, 1481, 1978.

53. Fung, K. W., Kuan, S. S., Sung, H. Y., and Guilbault, G. G., Methionine selective enzyme electrode, *Anal. Chem.*, 51, 2319, 1979.
54. Hikuma, M., Kubo, T., Yasuda, T., Karube, I., and Suzuki, S., Ammonia electrode with immobilized nitrifying bacteria, *Anal. Chem.*, 52, 1020, 1980.
55. Kjellén, K. G. and Neujahr, H. Y., Enzyme electrode for phenol, *Biotechnol. Bioeng.*, 22, 299, 1980.
56. Kulis, Y. Y. and Kadzyavskene, K. V., Bioelectrocatalysis. Lactate oxidizing electrode, *Doklady Akademii Nauk SSSR*, 239, 636, 1978.
57. Walters, R. R., Moriarty, B. E., and Buck, R. P., *Pseudomonas* bacterial electrode for determination of L-histidine, *Anal. Chem.*, 52, 1680, 1980.
58. Neujahr, H. Y., Enzyme probe for catechol, *Biotechnol. Bioeng.*, 22, 913, 1980.
59. Neujahr, H. Y. and Kjellén, K. G., Bioprobe electrode for phenol, *Biotechnol. Bioeng.*, 21, 671, 1979.
60. Wingard, L. B., Jr. and Millis, J. R., Cofactor reaction — diffusion kinetics for an alcohol dehydrogenase membrane, *Enzyme Eng.*, 5, 133, 1980.
61. Gorton, L., A Study of Modified Electrodes and Enzyme Reactors, Ph.D. thesis, Univ. of Lund, Sweden, 1981.
62. Jensen, M. A. and Rechnitz, G. A., Bacterial membrane electrode for L-cystein, *Anal. Chim. Acta*, 101, 125, 1978.
63. Paolantonio, C. L., Arnold, M. A., and Rechnitz, G. A., Serine-selective membrane probe based on immobilized anaerobic bacteria and a potentiometric ammonia gas sensor, *Anal. Chim. Acta*, 128, 121, 1981.
64. Karube, I., Matsunaga, T., Teraoka, N., and Suzuki, S., Microbioassay of phenylalanine in blood sera with a lactate electrode, *Anal. Chim. Acta*, 119, 271, 1980.
65. Riechel, T. L. and Rechnitz, G. A., Hybrid bacterial and enzyme membrane electrode with nicotinamide adenine denucleotide response, *J. Membr. Sci.*, 4, 243, 1978.
66. Wollenberger, U., Scheller, F., and Atrat, P., Microbial membrane electrode for the determination of cholesterol, *Anal. Lett.*, 13(B10), 825, 1980.
67. Wollenberger, U., Scheller, F., and Atrat, P., Microbial membrane electrode for steroid assay, *Anal. Lett.*, 13(B13), 1201, 1980.
68. Matsumoto, K., Seijo, H., Watanabe, T., Karube, I., Satoh, I., and Suzuki, S., Immobilized whole cell-based flow-type sensor for cephalosporins, *Anal. Chim. Acta*, 105, 429, 1979.
69. Kobos, R. K., Rice, D. J., and Flournoy, D. S., Bacterial membrane electrode for the determination of nitrate, *Anal. Chem.*, 51, 1122, 1979.
70. Mattiasson, B., Applications of immobilized whole cells in analysis, in *Immobilized Microbial Cells*, ACS Symp. Ser., Vol. 106, American Chemical Society, Washington, D.C., 1979, 203.
71. Daly, J. R., The application of quantitative cytochemistry to hormone bioassay, in *Recent Advances in Clinical Biochemistry*, Albertii, K. G. M. M., Ed., Churchill Livingstone, London, 1978, 129.
72. Peterson, A., Lewné, M., and Walum, E., Acute toxicity of organic solvents, heavy metals and DDT tested in cultures of mouse neuroblastona cells, *Toxicol. Lett.*, 9, 101, 1981.
73. Hikuma, M., Obana, H., Yasuda, T., Karube, I., and Suzuki, S., Amperometric determination of total assimilable sugars in fermentation broths with use of immobilized whole cells, *Enzyme Microb. Technol.*, 2, 234, 1980.
74. Grobler, S. R. and Rechnitz, G. A., Determination of D (+) glucose, D (+) mannose, D (+) galactose or D-fructose in a mixture of hexoses and pentoses by use of dental plaque coupled with a glass electrode, *Talanta*, 27, 283, 1980.
75. Mattiasson, B., unpublished.
76. Hikuma, M., Suzuki, H., Yasuda, T., Karube, I., and Suzuki, S., A rapid electrochemical method for assimilation test of microörganisms, *Eur. J. Appl. Microbiol. Biotechnol.*, 9, 305, 1980.
77. Karube, I., Matunaga, T., Mitsuda, S., and Suzuki, S., Microbial electrode BOD sensor, *Biotechnol. Bioeng.*, 19, 1535, 1977.
78. Karube, I., Matsunaga, T., and Suzuki, S., A new microbial electrode for BOD estimation, *J. Solid-Phase Biochem.*, 2, 97, 1977.
79. Karube, I., Mitsuda, S., Matunaga, T., and Suzuki, S., A rapid method for estimation of BOD by using immobilized microbial cells, *J. Ferment. Technol.*, 55, 243, 1977.
80. Hikuma, M., Suzuki, H., Yasuda, T., Karube, I., and Suzuki, S., Amperometric estimation of BOD by using living immobilized yeasts, *Eur. J. Appl. Microbiol. Biotechnol.*, 8, 289, 1979.
81. Buchnall, R. A., The use of cultured cells and tissues in the development of antiviral drugs, in *The Use of Alternatives in Drug Research*, Rowan, A. N. and Stratmann, C. J., Eds., Unwin Brothers Ltd., Old Woking, UK, 1980, 15.
82. Walum, E. and Peterson, A., Acute toxicity testing in cultures of mouse neuroblastoma cells, in Proceedings of the First CFN-Symposium, in press.

83. **Boyse, E. A., Old, L. J., and Chouroulinkov, I.**, Cytotoxic test for demonstration of mouse antibody, in *Methods in Medical Research,* Eisen, H. N., Ed., Vol. 10, Year-Book Medical Publishers, Chicago, 39, 1978.
84. **Lee, S., Robinson, F. M., and Wang, H. Y.**, Rapid determination of yeast viability, in *Third Symposium on Biotechnology in Energy Production and Conservation,* Scott, C. D., Ed., John Wiley & Sons, New York, 1981, 641.
85. **Karube, I., Matsunaga, T., Nakahara, T., Suzuki, S., and Kada, T.**, Preliminary screening of mutagens with a microbial sensor, *Anal. Chem.*, 53, 1024, 1981.
86. **Sakato, K., Tanaka, H., Samejima, H., Karube, I., and Suzuki, S.**, Electrochemical measurements of cell populations, *Ann. N.Y. Acad. Sci.*, 369, 321, 1981.
87. **Mattiasson, B. and Borrebaeck, C.**, Non-equilibrium, isokinetic enzyme immunoassay of insulin using reversibly immobilized antibodies, in *Enzyme Labeled Immunoassay of Hormones and Drugs,* Pal, S. B., Ed., Walter de Gruyter, Berlin, 1978, 91.
88. **Jacoby, B. and Wilchek, M.**, *Methods Enzymol.*, 34, 1974.
89. **Sharma, S. K. and Mahendroo, P. P.**, Affinity chromatography of cells and cell membranes, *J. Chrom.*, 184, 471, 1980.
90. **Danielsson, B., Mattiasson, B., Karlsson, R., and Winqvist, F.**, The use of enzyme thermistor in continuous measurements and enzyme reactor control, *Biotechnol. Bioeng.*, 21, 1749, 1979.
91. **Mandenius, C.-F., Danielsson, B., and Mattiasson, B.**, Enzyme thermistor control of the substrate concentration at a fermentation with immobilized yeast, *Acta Chem. Scand.*, B34, 463, 1980.
92. **Mandenius, C.-F., Danielsson, B., and Mattiasson, B.**, Process control of an ethanol fermentation with an enzyme thermistor as a sucrose sensor, *Biotechnol. Lett.*, 3, 629, 1981.

Chapter 7

CELL MEMBRANES AS ADSORBENTS IN AFFINITY CHROMATOGRAPHY

Jose-Luis Ochoa

TABLE OF CONTENTS

I. INTRODUCTION

Many substances present in biological systems possess the remarkable ability to interact in a more or less specific fashion with other molecules, e.g., enzymes, antibodies, hormones, lectins, etc. For their purification, this fact has led to the development of a very useful technique which is principally based on the biospecific recognition of a particular substance for another, namely affinity chromatography.[1,2] The application of affinity chromatography as a separation tool implies the immobilization of a particular compound on a support or matrix.[1,2] The matrix should not possess any undesirable nonspecific adsorption properties by itself and the immobilized substance (ligand) should retain fully its capacity to interact biospecifically with the molecule of interest (ligant). Very sharp biospecific adsorbents, however, are not always desirable since their cost increases as a result of a decreased versatility and reduced potential in group separation. That is, more 'general ligands', such as nucleotide cofactors, complex carbohydrates, or even macromolecules may be preferred as they make possible the fractionation of a number of substances from different sources and, possibly, with different biological activities.[3-6] In the purification of cell membranes interacting compounds, the membrane itself may serve as a suitable and economic multispecific affinity adsorbent. It is apparent that several of the most important biological reactions occur at the cell membrane level. Therefore, it seems reasonable to believe that cell membrane residues may also be useful in the isolation of a number of interesting biologically active molecules.

II. METHODS FOR CELL MEMBRANE PREPARATION

Being aware of the fact that practically any tissue from animal or plant origin may serve as source for cell membrane adsorbents, our discussion is limited to the preparation of erythrocyte ghosts. The best argument for doing so is that red cells are easily available and have been used for studying the cell membrane structure and function. Furthermore, erythrocyte stroma is becoming a popular adsorbent for the isolation of a number of biological molecules.[22-29]

There are as many different types of red cell ghosts as there are ways of causing hemolysis.[7] The most common, however, is the hypotonic disruption which results in the preparation of ghosts with similar composition to that of the membrane of the intact cell.[8,9] Very recently, a filtration step that avoids the necessity of repeated centrifugation-washing steps for the preparation of hemoglobin-free ghost was introduced,[10-11] thus permitting the treatment of considerably larger amounts of red cells in short periods. The filtration can be carried out either by molecular-sieve chromatography[10] or by passing the suspension through an apparatus equipped with a filter that eliminates the hemoglobin while retaining the ghost membranes.[11] The quality of the membranes obtained by either method, as judged by SDS-electrophoretic analysis, is comparable to that resulting from repeated centrifugation-washing steps.[8-9]

The purity of ghost preparations is usually assumed to be inversely related to its hemoglobin content, since hemoglobin is not considered an essential red cell membrane constituent.[8,9] However, at least one other membrane protein, the enzyme *p*-nitrophenylphosphatase, has been shown to follow a similar release pattern to that of hemoglobin.[12]

Not all the membrane proteins are released with equal ease. Osmolarity, ionic strength, pH temperature, and even type of salt and other added agents (such as EDTA, NaN_3, etc.), determine the nature of the ghosts prepared.[7,13,14] Spectrin, for instance, is rapidly lost into the aqueous media at very low ionic strength. Glycophorin, on the other hand, is not released into aqueous solutions under such conditions.[15,16]

In general, if the ionic concentration is increased to values beyond 1 mM, protein losses decrease and reach a minimum at 10 mM. At higher concentrations the release of proteins increases again and may be accompanied by some loss of membrane lipids.[9,17] The characteristic pattern of protein extraction is possibly related to electrostatic interactions between the inorganic ions and the membrane proteins as follows: on one hand, at low ionic strength the electrostatic repulsion between the membrane proteins can be expected to increase.[18] This could facilitate the release of proteins. At high ionic strength a salting-in effect[19] may enhance the solubilization. The minimum of solubility could represent the balancing out of the two effects.[7]

As pointed out, in spite of the exposure of the cell membrane to low ionic strength and the expected rearrangements of the membrane components, the ghosts of erythrocytes usually have the same outer appearance as intact erythrocytes.[8,9] Although the ghosts may show a higher permeability for small and large ions and molecules,[12,20,21] it has been assumed that they possess the same molecular organization as the intact membrane and therefore they have been used not only as membrane models but also to study the biological phenomena occurring at the cell membrane level. As far as their utilization as an affinity adsorbent is concerned, the purification of several membrane interacting proteins by this means[22-29] proves that at least some of the cell membrane receptors retain their capacity of interacting biospecifically with such molecules. The stability of the cell membrane can be improved by treatment with glutaraldehyde or formaldehyde.[22,23] Copolymers of stroma with other inert proteins such as albumin can also be advantageous[22] since this might facilitate the preparation of coarse particles for column operation.

III. THE RED CELL MEMBRANE AS AN AFFINITY ADSORBENT

In the previous sections, it has been suggested that the erythrocyte ghosts, obtained by hypotonic lysis of red cells, and the intact red cell membrane, are in appearance, very much alike. It might be expected, therefore, that the ghosts, or stroma, should retain also the capacity of the red cell to interact biospecifically with a number of biologically active molecules such as lectins, antibodies, etc. To illustrate this, Table 1 shows the results obtained by various authors who used the formalinized erythrocyte and stroma as an affinity adsorbent for the isolation of lectins.

It should be mentioned, however, that comparisons within the results shown in Table 1 are not possible, since the experimental conditions were different in all cases. In addition, as indicated in Table 1, erythrocytes of various sources (pig, human, chicken, rabbit, and sheep) were employed in each case, and therefore, differences in relative adsorption capacity might also be expected. For these reasons Table 1 serves uniquely to indicate that both formalinized whole red cells, and/or red cell membrane residues, are useful adsorbents in the isolation of biologically active molecules such as lectins.

Different alternatives have been proposed as to how to use the stroma or the whole red blood cells in affinity chromatography. It is obvious that the column procedure[24,25,28,29] is rather more advantageous than the batch one[35-36] as the loading and elution steps are more easily carried out. Further, a significant saving on time when handling large samples, a higher adsorption capacity, as well as a better reproducibility, are common characteristics of the chromatography in columns.

Since lectins are carbohydrate binding proteins,[30,31] their elution from stroma, or whole red blood cells columns is done by sugar solutions.[23,26-29] Very often, however, drastic conditions have to be used to obtain adequate yields or recoveries.[22,24,25] Low pH as well as washings with hypotonic solutions preclude the possibility of using the whole red cell as a suitable adsorbent. In such cases, stroma is preferred.

Table 1
THE PURIFICATION OF LECTINS USING WHOLE RED BLOOD CELLS (RBC) AND/OR ERYTHROCYTE GHOSTS AS ADSORBENTS

	mg of crude extract added		mg of purified lectin recovered		
	RBC	Ghost	RBC	Ghost	Ref.
Lectin					
Concanavalin-A	9	740	1.7	8.4	22,23
Wheat germ agglutinin	84	50	2.9	0.9	22,23
Soy bean agglutinin	—	25	—	2.1	22
Ulex europeus	64	—	1.8	—	23
Lima bean	300	—	2.0	—	23
Ph. vulgaris	—	60	—	5	24
Ph. coccineus	—	102	—	18.8	24,25
Potato fruit	1,120	—	1.3	—	26
Limulus polyphemus	8,442	—	4.0	—	27

The immobilization of stroma, or whole red blood cells, has been approached in various ways: the covalent immobilization is not as popular as the physical entrapment technique simply, because the latter is just as efficient and more easily carried out.

When using whole cells, however, some important factors should be taken into account:[42]

1. The mixing of the cells and the polymer solution should be easy and the mixture should be stable at liquid state.
2. The mixture should be induced to the gel state from the liquid state under mild conditions which do not modify the structure of the membrane components.
3. Formed gels should possess good mechanical resistance and be stable at high temperatures (60 to 70°C) over a wide pH range.
4. The pore size of this gel matrix should be small enough so that the cells do not leak out of the gel while the molecules to be isolated can diffuse freely through the gel lattice.

As pointed out, other cells may as well be used as affinity adsorbents for the separation of biologically active molecules. Yeasts, for example, have been physically trapped in polyacrylamide beads and used for the isolation of *Ricinus communis* L. agglutinin.[33] This immobilized yeast, (*C. lipolytica*), showed a lower yield, or adsorption capacity, as compared to similar columns containing stroma or an adsorbent for which *R. communis* lectin possesses affinity (Sepharose 4B) (Table 2). The difference in yield has been ascribed to the density of binding sites that apparently are less numerous on the surface of the yeast cells than in stroma or in the polysaccharide network of Sepharose 4B.

Bacterial immunoadsorbents consisting of organisms dispersed in agar beads have been also very useful for the removal of cross-reacting agglutinins.[34] Electrostatically immobilized bacterial cells on ion exchangers[35] (cellex-T; triethylaminoethyl-cellulose, DEAE-cellulose, and Dowex) have also been successfully applied for the specific removal of antibodies from whole serum.[36] In this instance, anion exchange fractionation of the whole serum by the ion-exchanger matrix itself is prevented by using buffers of relatively high ionic strength (0.25 *M* NaCl). Since elution is necessarily carried out at low pH (2.3), the bacterial cells are pretreated with formalin and washed with acetone to ensure that they remain in the column.

Table 2
THE PURIFICATION OF *RICINUS COMMUNIS* LECTIN ON DIFFERENT ADSORBENTS

	Yield[a]
Stroma	0.28
Candida lipolytica	0.08
Sepharose 4B	0.21

[a] Protein yield is expressed as the ratio of the optical densities (O.D.) of the eluate and of the extract passed through the column.[33]

Bacteria can also be polymerized and mixed with an inert support to serve as affinity adsorbents.[37-39] Since the ratio bacteria:inert support can be varied, adsorbents of different capacities are easy to prepare by this manner.

IV. OTHER CELL ORGANELLES

The idea of using immobilization of cell organelles for a number of purposes in affinity chromatography has been attempted in a number of ways.[40] As in the immobilization of whole microbial cells, the use of organelles eliminates the often tedious, time-consuming, and expensive steps involved in purifying intracellular enzymes or receptors. In addition, the stability of enzymes used for continuous operation is improved when surrounded by its natural environment and, further, special advantages are also gained if the processes are requiring multienzyme systems and/or cofactors.

In the cell, organelles represent highly ordered compartments where specific reaction sequences are carried out. Unlike the intact cells, isolated organelles are not protected by a thick cell wall and are therefore very fragile. Organelle immobilization thus presents additional problems. Special care should be taken to preserve the activities of the individual enzymes, or receptors, their structural and conformational integrity. Some success has been achieved in the immobilization and stabilization of spinach leaf chloroplasts.[40]

Intact nucleic and nuclear matrices possess also the ability of binding lectins in a specific manner.[41-42] Ribosomes, on the other hand, seem to owe their lectin binding capacity not to its own components but rather to glycoproteins firmly adhered to its surface.[43] Not only lectins and antibodies are capable of binding to cell membranes; polypeptide hormones, for example, are known to bind to receptor glycoproteins located in the surface of target cells provoking the regulations of cell surface functions such as glucose and amino acid transport.[44] One such group of polypeptide hormones includes glucagon, ACTH, and thyrotropin which activate the adenylate cyclase enzyme complex in the plasma membrane of target cells giving origin to a second intracellular messenger cAMP that induce many, if not all, of the actions of the hormones.[45] The possibility of purifying this kind of substances by utilizing the cell membrane as an affinity adsorbent, however, remains yet to be explored.

In conclusion, the use of cell membranes, and/or cell organelles, as natural biospecific adsorbents in affinity chromatography awaits further development. The researchers should look at this field as an interesting and very promising alternative for the purification of a large variety of biologically active compounds.

REFERENCES

1. Jakobi, W. B. and Wilchek, M., Eds., *Affinity Techniques, Methods in Enzymology,* Academic Press, New York, 1974.
2. Lowe, C., *An Introduction to Affinity Chromatography,* Work, T. S. and Work, E., Eds., North Holland, Amsterdam, 1979.
3. Mosbach, K., Separation of isoenzymes, *Methods Enzymol.,* 34, 595, 1974.
4. Sela, B. A., Wang, J. L., and Edelman, G. M., Isolation of lectins of different specificities on a single affinity adsorbent, *J. Biol. Chem.,* 250, 7535, 1975.
5. Ratcliffe, R. M., Baker, D. A., and Lemieux, R. U., Synthesis of the (-D-Gal-(1-3)- -D-GalNAc)-Antigenic determinant in a form useful for the preparation of an effective artificial antigen and the corresponding immunoadsorbent, *Carbohydrate Res.,* 93, 35, 1981.
6. Pazur, J. H., Forry, K. R., Tominaga, Y., and Ball, E. M., Anti-glycosyl antibodies: antibodies directed against the carbohydrate moeities of a glycoprotein, *Biochem. Biophys. Res. Commun.,* 100, 420, 1981.
7. Schwoch, G. and Passow, H., Preparation and properties of human erythrocyte ghosts, *Mol. Cell. Biochem.,* 2, 197, 1973.
8. Hillier, J. and Hoffman, J. F., The ultrastructure of the plasma membrane as determined by electron microscope, *J. Cell. Comp. Physiol.,* 42, 203, 1953.
9. Dodge, J. T., Mitchell, C., and Hanahan, D. J., Preparation and chemical characteristics of hemoglobin-free ghosts of human erythrocytes, *Arch. Biochem. Biophys.,* 100, 119, 1962.
10. Froman, G., Acevedo, F., and Hjertén, S., A molecular sieving method for preparing erythrocyte membranes, *Prep. Biochem.,* 10, 59, 1980.
11. Rosenberry, T. L., Chen, J. F., Lee, M. M. L., Moulton, T. A., and Onigman, P., The large scale isolation of human erythrocyte membranes by high volume molecular filtration, *J. Biochem. Biophys. Meth.,* 4, 39, 1981.
12. Bramley, T. A., Coleman, R., and Finean, J. B., Chemical enzymological and permeability properties of human erythrocyte ghosts prepared by hypotonic lysis in media of different osmolarities, *Biochim. Biophys. Acta,* 244, 752, 1971.
13. Glaser, R., Svetina, S., Brumen, M., Learche, D., Quemada, D., Mills, P., and Dufaux, J., Stationary ionic states of human red blood cells consequences for planning and evaluation of experiments, *Stud. Biophys.,* 81, 43, 1980.
14. Kant, J. A. and Steck, T. L., Cation-impermeable inside-out and right-side out vesicles from human erythrocyte membranes, *Nature (London) Biol.,* 240, 26, 1972.
15. Marchesi, V. T. and Steers, E., Selective solubilization of a protein component of the red cell membrane, *Science,* 159, 203, 1968.
16. Tanner, M. J. A. and Boxer, D. H., Separation and some properties of the major proteins of the human erythrocyte membrane, *Biochem. J.,* 129, 333, 1972.
17. Mitchell, C. D. and Hanahan, D. J., Solubilization of certain proteins from the human erythrocyte stroma, *Biochemistry J.,* 5, 51, 1966.
18. Pollack, W. and Reckel, R. P., A reappraisal of the forces involved in hemagglutination, *Int. Arch. Allergy Appl. Immunol.,* 54, 29, 1977.
19. Hatefi, Y. and Hastein, W. G., Solubilization of particulate proteins and nonelectrolytes by chaotropic agents, *Biochemistry,* 62, 1129, 1969.
20. Triplett, R. B. and Carraway, K. L., Proteolytic digestion of erythrocytes, resealed ghosts and isolated membranes, *Biochemistry,* 11, 2897, 1972.
21. Philips, D. R. and Morrison, M., Exposed protein on the intact human erythrocyte, *Biochemistry,* 10, 1766, 6971.
22. Avrameas, S. and Guilbert, B., Biologically active water-insoluble protein polymers. Their use for the isolation of specifically interacting proteins, *Biochimie,* 53, 603, 1971.
23. Reitherman, R. W., Rosen, S. D., and Barondes, S. H., Lectin purification using formalinised erythrocytes as a general affinity adsorbent, *Nature (London),* 248, 599, 1974.
24. Ochoa, J. L. and Kristiansen, T., Stroma: as an affinity adsorbent for non-inhabitable lectins, *FEBS Lett.,* 90, 145, 1978.
25. Ochoa, J. L. and Kristiansen, T., Purification and partial characterization of *Phaseolus coccineus* var. 'alubia' agglutinin, *Biochim. Biophys. Acta,* 705, 396, 1982.
26. Kilpatrick, D. C., Isolation of a lectin from the pericarp potato *(Solanum tuberosum)* fruits, *Biochem. J.,* 191, 273, 1980.
27. Nowak, T. P. and Barondes, S. H., Agglutinin from *Limulus polyphemus.* Purification with formalinized Horse erythrocytes as the affinity adsorbent, *Biochim. Biophys. Acta,* 393, 115, 1975.
28. Betail, G., Coulet, M., Guillot, J., and Scandariato, M., Emploi de colonnes d'hematies pour la purification de quelques agglutinines vegetales, *C. R. Seances Soc. Biol.,* 165, 1304, 1971.

29. **Betail, G., Coulet, M., Genaud, L., Guillot, J., and Scandariato, M.**, Les stromas erythrocytaires inclus en gel de polyacrylamide. Applications a la chromatographie d'affinite, *C. R. Seances Soc. Biol.*, 169, 561, 1975.
30. **Goldstein, I. J. and Hayes, C. E.**, The lectins: carbohydrate-binding proteins of plants and animals, *Adv. Carbohydr. Chem. Biochem.*, 35, 127, 1978.
31. **Lis, H. and Sharon, N.**, Lectins: their chemistry and application to immunology, in *The Antigens*, Vol. 4, Sela, M., Ed., Academic Press, New York, 429, 1977.
32. **Genaud, L., Guillet, J., Damez, M., and Coulet, M.**, Use of Yeasts in affinity chromatography, *J. Immunol. Meth.*, 22, 339, 1978.
33. **Thomson, R. O., Walker, P. D., and Harris, E.**, Bacterial immunoadsorbent columns, *Appl. Microbiol.*, 18, 874, 1969.
34. **McKinney, R. M., Thacker, L., Wong, M. C., and Hebert, G. A.**, Improved immunoadsorption procedure with anion-exchange bacterial cell columns, *J. Immunol. Meth.*, 21, 1, 1978.
35. **Weetall, H. H.**, Immunoadsorbent for the isolation of bacterial specific antibodies, *J. Bacteriol.*, 93, 1876, 1967.
36. **Fine, D. H., Tabak, L., and Stevens, R.**, Affinity chromatography of antiserum to a gram negative organism, *J. Immunol. Meth.*, 16, 91, 1977.
37. **Morrison, R. B.**, A new method for agglutinin adsorption, *Lancet*, 2, 1288, 1961.
38. **Cope, J. B., Redys, J. J., Hibbard, E. W., and Borman, E. K.**, Improved method of adsorbing immunofluorescence reagents, *Health Lab. Sci.*, 5, 208.
39. **Venkatasubramanian, K.**, Applications of enzyme technology in biosaline environments, *J. Solid-Phase*, 2, 305, 1977.
40. **Kayano, H., Matsunaga, T., Karube, I., and Suzuki, S.**, Photochemical energy conversion system using immobilized chloroplasts, *Biotechnol. Bioeng.*, 22, 2283, 1981.
41. **Sevaljevic, L., Poznanovic, G., Petrovic, M., and Krtolica, K.**, Protein kinase activity and lectin binding ability to the nuclear matrix, *Biochem. Int.*, 2, 77, 1981.
42. **Monneron, A. and Segretain, D.**, Extensive binding of concanavalin A to the nuclear membrane, *FEBS Lett.*, 42, 209, 1974.
43. **Michel, S., Madjar, J. J., Reboud, J. P., and Cozzone, A. J.**, Interaction of concanavalin A with individual proteins from bacterial and mammalian ribosomes, *Int. J. Biochem.*, 13, 141, 1981.
44. **Goldfine, I. D.**, Interaction of insulin, polypeptide hormones and growth factors with intracellular membranes, *Biochem. Biophys. Acta*, 650, 53, 1981.
45. **Goldfine, I. D.**, Insulin receptors and the site of action of insulin, *Life Sci.*, 23, 2639, 1978.

Chapter 8

PROCESS ENGINEERING CONSIDERATIONS IN THE DEVELOPMENT OF IMMOBILIZED LIVING CELL SYSTEMS

K. Venkatasubramanian and S. B. Karkare

TABLE OF CONTENTS

I. INTRODUCTION

Continuous biocatalysis by immobilized microbial cells represents a relatively new approach to replace conventional bioconversion processes. Immobilized cell systems mediating simple, single enzyme reactions have already been reduced to commercial practice. Production of high fructose corn syrup (HFCS) is a successful example of this approach. However, the evolution of the next generation immobilized cell reactors to carry out complex fermentation processes — characterized by multiple sequential reaction pathways involving coenzymes — is still in its infancy.

The advent of recombinant DNA technology in the past few years has added a dramatically new dimension to the development of biotechnology. While it is possible to utilize the genetically engineered organisms with conventional fermentation technology, a parallel engineering development of new biochemical process technology is necessary in order to fully harness their potential. Immobilized microbial cell systems could form the backbone of such a new process technology. Recognition of this approach is clearly evident from even a cursory glance at the recent literature. Several excellent reviews[1-6] summarize the different methods of cell immobilization and their applications.

Many of the reports in the literature are concerned with immobilization technique, activity of the cells in the immobilized form, and at best a rudimentary treatment of the reaction kinetics. Process engineering aspects of immobilized reactor systems are often not adequately discussed in these papers. This is even more so in the case of immobilized living microorganisms.

Many researchers have demonstrated convincingly that a microorganism must be maintained in the living state, even in the immobilized form in order to effect complex biochemical reaction pathways.[6,7] This is not entirely surprising from a cellular bioenergetic point of view, as the cells would have to transform the chemical energy contained in the substrate to biologically useful energy forms. There is considerable evidence that adsorbed and gel-entrapped microorganisms can replicate within the gel matrix.[8,9] This approach holds considerable promise for the biosynthesis of growth associated products by immobilized cells, as well as for the rejuvenation of catalytic activity in other cases. Since this concept is still in its early stage of development, little attention has been paid to the process engineering of reactor systems embodying such live biocatalysts. In particular, the role of cellular physiology and bioenergetics, as they relate to overall reactor design and operation, has not been investigated systematically. These are the key points addressed in this chapter. Our purpose here is not to review everything there is on hand, but rather to analyze the recent developments dealing with these facets of immobilized cell systems. Based on this, we shall outline several considerations for further process development work. Since this entire field is still quite young, the emphasis of this chapter is prospective rather than retrospective in nature.

II. GENERAL PROCESS ENGINEERING CONSIDERATIONS

For the efficient development of any bioreactor system employing immobilized living cells (IMC), a series of *a priori* process engineering considerations are apparent. The choice of the carrier matrix and the immobilization method would greatly influence the expressed catalytic activity of the immobilized cell. While highly porous supports would provide a large surface area, they are likely to encounter considerable mass transfer resistances. Similarly, the immobilization technique would determine whether or not the cells could be retained in the living state.

From a practical standpoint, the catalytic activity and the operational stability of the immobilized cells would dictate the effective cost of the catalyst for a given bioconversion process. The possibility that the live cells could be rejuvenated periodically adds an exciting dimension here. However, one needs to ascertain how many such rejuvenation cycles that the system can be subjected to (see Chapter 2, this volume). Whether or not the intrinsic specific activity of the cell could be maintained without any loss in the immobilized state is also an interesting question in this connection. It is likely that the physiological state of the organism would play a critical role in determining this.

In addition to the catalytic activity, the volumetric productivity of the reactor system is also important, since it would establish the reactor size for a given throughput level. However, the volumetric productivity question should be addressed in conjunction with the product concentration obtainable in the immobilized cell system. The effective product concentration in the reactor exit would dictate the extent and cost of downstream processing. This is often an ignored element in the development of bioreactor systems.

Since many fermentation processes are substrate cost intensive, the extent of substrate conversion and product yield are critical factors governing the operating cost. The ability of the immobilized cell systems to handle commercial substrates — which are often particulate or colloidal in nature — is an important factor in this regard.

Since rapidly utilizable substrates are expected to be used for many of the processes in question, it is essential to operate the reactor system without microbial contamination.

Transport constraints become even more important in the case of immobilized living cell systems. The problem is further exacerbated for aerobic organisms for which oxygen supply may very well become the limiting factor. If active transport mechanisms are involved, cellular energetics becomes even more critical. The immobilization matrix could also facilitate or retard the transport characteristics of the system.

Another important consideration is the design of a suitable reactor configuration. Since the cells must be maintained in the living state, some of the commonly employed reactor types such as the packed bed reactor may not be readily applicable. Development of proper scale-up procedures is still an unexplored territory. Operational control of the reactor systems to optimize production levels and conversion efficiency is also a relevant consideration. Finally, careful analysis of the overall process economics of immobilized cell systems is needed to identify critical sensitivities which bear upon the commercial viability of the immobilized cell approach.

The following sections elaborate these points.

III. CARRIER MATRIX AND IMMOBILIZATION METHOD

A variety of support matrices and immobilization techniques are available for the immobilization of microbial cells. Some are more suitable for the retention of the immobilized cells in their living state. For instance, cells entrapped within gels of different hydrocolloids (e.g., carrageenan and alginates) provide for a gentle method of immobilization. Since these gels are also highly hydrated, the mass transfer problem is also diminished somewhat. Both inorganic and organic carrier matrices have been used as supports. The former can be made to provide large surface areas per unit volume/weight of the carrier.

Messing[10,11] characterizes a number of such matrices in terms of their pore size and available surface area. In fact, he has estimated the average pore size needed for the initial entrapment and further propagation of different types of microbial cells inside these pores. The difficulty that one is likely to encounter with highly porous beads is the inability to supply adequate amounts of nutrients and oxygen to those organisms residing deep inside the pores. Thus, a balance must be struck between the surface area needed and the ability to achieve efficient mass transfer.

Certain carrier matrices appear to promote autolysis of the cells. A case in point here is described by Chibata[12] for polyacrylamide gels.

Another immobilization approach that appears to be successful is the physical sorption of the cells onto the surface of different resins. If adequate cell loading could be achieved, this would be an attractive immobilization method. Some elution of the cells into the medium is perhaps unavoidable. However, this may not necessarily be undesirable as we shall see later. The adhesion characteristics of the cells in question would be the governing factor in the applicability of this method. It is worth noting that both the adsorption and entrapment methods lend themselves more readily for easy scale-up.

Entrapment of cells inside hollow fibers provides an interesting alternative (see Chapter 5, Volume I). It is possible to increase the effective cell density several times when compared to free cell fermentations. Yet another promising approach involves the use of two phase systems[13] (see Chapter 7, Volume I).

The use of chemical reagents to cross-link the matrix or to covalently link the cells to a carrier surface often leads to the destruction of some of the catalytic activity. For instance, our attempts to strengthen the carrageenan matrix by glutaraldehyde cross-linking resulted in reduced activity for the case of immobilization of *Acetobacter suboxydans.*[14] A correlation between the extent of glutaraldehyde treatment and loss of activity has been demonstrated.[15,16]

Only a few attempts have been made to examine the distribution of the immobilized cells in the matrix. In the case of adsorption, the cells appear to distribute themselves uniformly throughout the matrix. On the other hand, there appears to be a preferential clustering of the cells in the cavities of gel matrices. Uniform distribution of the cells and their activity would be an important factor to achieve good contact efficiency in reactor systems.

Another point that must be considered in the choice of the immobilization method is the ability to accomplish it under sterile conditions. We have developed a procedure to successfully entrap cells *in situ* in a fermentor in carrageenan gels under sterile conditions. This consideration becomes even more important in the case of higher organisms.

IV. ACTIVITY AND PRODUCTIVITY

In the case of immobilized enzymes and nonliving cells, it has been customary to express the activity of the immobilized system in terms of activity units (e.g., International Units per unit weight of the catalyst). This approach would be valid if the number and viability of the cells on the matrix remain constant. Since it is not likely to be so for fixed living cells, it is more useful to employ a volumetric productivity or production rate measure to characterize the immobilized catalyst. This can be expressed, for instance, in grams product/liter reactor volume/hour. (To be even more precise, the product concentration should also be stated along with the volumetric productivity.) This would also provide a ready comparison with continuous culture systems.

As pointed out earlier, volumetric productivity is the key factor in establishing the size of the reactor for a desired throughput level.

One of the inherent advantages of the immobilized cell approach is the dramatic increase in volumetric productivity over conventional batch fermentations. For example, ethanol productivity has been shown to increase by as much as 40 to 50 times.[17] Such increases are attributable mainly to (1) the increased (volumetric) cell density in the immobilized cell reactor, and (2) the ability to operate the reactor at very high flow rates without the danger of a culture washout as is often encountered in continuous culture systems.

Another question which has not been fully explored is the intrinsic catalytic activity of the individual cell in the immobilized state. The volumetric productivity represents a composite or aggregate behavior of the total cell population, and it does not reflect the catalytic performance of the individual cell. The literature data on this point appear to be quite mixed. While there is some indication that the intrinsic activity decreases upon immobilization in several cases, it also seems to register an increase in some instances. It is reasonable to speculate that the physiology of the organisms in the immobilized state undergoes a change leading to the observed differences in the intrinsic activity. However, it is not clear whether or not the physical stress that the organisms are subjected to upon immobilization contributes to such a physiological change. In a recent paper, Bandyopadhyay and Ghose[15] present some electron microscopic evidence to demonstrate the physical stress on the organism in the immobilized state. In this particular case (yeast immobilization), they have observed a significant reduction in the generation time of the immobilized cells and attribute this to a physiological change. It is obvious that this aspect has to be carefully investigated, for it provides another vehicle to improve the overall performance of the immobilized cell systems.

V. CATALYST LIFE

The decrease in catalytic activity with operating time is generally characterized by a catalyst half-life in the case of immobilized enzymes and the nonliving cells. These biocatalysts are usually loaded into a column and then left there until the activity is exhausted down to a predetermined level — typically three half-lives in industrial operations. Hence, this can be described as a "static" biocatalytic system.

It is possible to rejuvenate the activity of the immobilized living cells. In the most idealized version, it is conceivable that a constant volumetric productivity level (g product/ℓ reactor vol/hr) can be maintained over long periods of time. Thus, it is not entirely appropriate to invoke the concept of catalyst half-life to describe the "dynamic" state achievable with bound living cells. The more relevant question here is how many rejuvenation cycles could one achieve in continuous operation.

It is possible to maintain constant volumetric productivity level by manipulating a number of process variables. For example, the nutrient mix fed to the reactor could be so modulated as to favor either cell growth or product formation, especially in the case of nongrowth associated products. We propose the term "metabolic switch" to describe this. The switch here signifies a change in the composition of the nutrient medium. We have demonstrated that it is possible to promote the biosynthesis of an antibiotic, candicidin, by immobilized *Streptomyces griseus* by eliminating phosphate in the medium. This is really an extrapolation of the work done with free cells.[18] Similarly, modulation of nitrogen level could be used to control citric acid production.[16]

As opposed to the secondary metabolites, the nutrient addition strategy would be quite different for the synthesis of growth associated products. Here one would expect to use the same nutrient medium to achieve growth and primary metabolite production simultaneously. Under these conditions, cell growth would probably occur in both the attached and the bulk phases. We have shown this to be the case for the oxidation of ethanol to acetic acid which is a highly growth associated process. A constant volumetric productivity level in this case could be achieved by maintaining a dynamic balance between the cells in the two phases. In such cases, the production of biomass at enhanced rates result from cell growth in both the immobilized as well as the bulk phase. Therefore, it would seem appropriate to characterize such reactors "hybrid reactors". Thus, the hybrid reactors represent a combination of the concepts of cell immobilization and continuous culturing of microorganisms. The mathematical basis for increased biomass generation obtainable in the hybrid reactors has been formulated and set forth recently.[19]

VI. PRODUCT YIELD

The efficient conversion of substrate to product is especially critical for complex catalytic processes. Substrate costs are often intensive, particularly for commodity-like products (e.g., ethanol) where substrate cost could form as much as 30 to 50% of the total production cost. Therefore, commercial acceptability of IMC systems cannot be achieved unless the product yields are close to those achievable in conventional batch fermentations. This point is seldom addressed in many of the publications.

It is conceivable that appropriate biocatalytic reactor design and operating strategies could be evolved to tackle this problem. Partial recycle of unconverted substrate and sequential operation in a reactor series are but two possible approaches — provided they are cost efficient.

Byproduct formation is also a factor contributing to the overall substrate conversion efficiency. If side reactions are not properly controlled, they reduce the yield of the desired product besides complicating the product recovery/purification scheme. Selection of a suitable microbial strain with little or no propensity to byproduct formation and preservation of the strain in the correct physiology in the immobilized state should help alleviate this problem. In our work on citric acid synthesis by immobilized cells, chromatographic analysis of the reaction products showed the presence of isocitric acid as a byproduct, but not at a level significantly more than that observed in the corresponding free cell fermentation.[7]

Product yield would also be affected by the maintenance energy requirements of the cell in the immobilized state. There are contradictory evidences/claims in the literature on the relative maintenance energy requirements of the microbial cells in the free vs. the immobilized state. Perhaps the larger question which underscores this point is the relative distribution of the substrate energy between cell growth, product formation, and maintenance. It is meritoriously attractive to speculate that immobilized cell systems offer the possibility of channeling a large portion of the substrate into product formation at the expense of the other two energy consuming components. This point has simply not been investigated systematically. We are now in the process of performing total carbon and nitrogen balances in the free and the immobilized states as a means of investigating this issue. In passing, it must be mentioned that if active or facilitated transport mechanisms are to be operable in the immobilized state, it would introduce an additional complexity in assessing the maintenance energy question.

VII. MASS TRANSFER CONSIDERATIONS

Just as in the case of immobilized single enzyme and nonliving cell systems, the role of mass transfer in the overall heterogeneous catalysis can be classified broadly into two categories: external (film) mass transport and internal diffusion. While a fairly good deal of work has been published in the literature on these effects for simple immobilized enzyme systems, little information is available for immobilized living cell mediated processes.

The mathematical treatment of the substrate flow up to the surface of the immobilized cell would be essentially the same as that for the above case. In addition, we have to consider here the diffusion of the substrate and the product through the barrier imposed by the cell envelope (i.e., cell wall and cell membrane) itself.

It can generally be assumed with considerable certainty that the substrate transport into the microbial cell is characterized by passive diffusion in the case of nonliving immobilized cells. The precise mechanism of substrate transport into an immobilized living cell has simply not been studied yet. However, it is reasonable to suggest that active or facilitated transport mechanisms would prevail at least for those substrate molecules which are transported in such manner in equivalent submerged fermentations. Here again, maintenance of the appropriate physiological state of the organism in the immobilized state can be surmised to be critical in achieving this. Thus, the type of diffusion mechanism would depend on the organism and true limiting substrate itself.

There have been many published reports which suggest nutrient depletion as the limiting factor in truly making use of the internal pores of an immobilization matrix. In fact, in many instances, the live cells tend to concentrate on or near the carrier surface. Because of this, surface properties of the matrix could be far more important in the case of fixed living cells.

Another important transport problem relates to oxygen transfer into the immobilized cells in the case of aerobic organisms. A corresponding problem is carbon dioxide removal from the immobilized cell reactor (see also Chapter 3, this volume). Adequate oxygen supply can be ensured in traditional fermentating through the use of properly designed aeration and agitation equipment. It may not be feasible to extrapolate this approach to immobilized cell reactors. Conversely, it may be possible to enrich air with oxygen since the reactors are much smaller in scale and better designed. Fluidized bed configurations can also be employed to improve gas-liquid-solid contact. As yet, these design alternatives have not been carefully evaluated.

VIII. REACTOR TYPES

Several reactor configurations have been employed for carrying out simple bioconversion processes, i.e., with immobilized enzymes and nonliving cells. These include packed bed reactors, continuous flow stirred tank reactors, fluidized bed reactors, and hollow fiber reactors. The relative advantages and disadvantages of the different reactor types, especially as they relate to complex processes carried out by immobilized living cells, are discussed in detail elsewhere.[19] They are summarized in a succinct tabular form here (Table 1).

Fluidized bed reactors and hollow fiber reactors appear to offer some unique advantages for the case of living immobilized cells. In addition to facilitating the oxygen transfer-CO_2 removal problem, the fluidized bed reactor provides better solid-fluid mixing characteristics. It may also be possible to achieve a stable cell population more readily since particles with lower cell loading move continuously from substrate-depleted zones to high substrate zones and vice versa by virtue of their changing densities. This must be contrasted against (1) the difficulty to fluidize the immobilized particles, and (2) the lack of reliable procedures for scale-up and operation.

IX. PRACTICAL CONSIDERATIONS IN REACTOR DESIGN AND OPERATION

Other publications from our laboratory[20,21] outline many of the practical considerations involved in the design and operation of bound cell reactors, especially as they relate to simple bioconversions. The principles discussed therein are generally applicable to more complex immobilized living cells as well. Due to the additional complexity of live cell systems, many other considerations also come into play.

Much work remains to be done in the rigorous mathematical analysis of IMC reactors. The autocatalytic cell growth reaction should be coupled to the bioconversion step. Further, the mass transfer steps should be superimposed on this. A recent publication from our laboratory describes the unique problems of reactor design and scale-up for IMC systems.[19]

Table 1
CHARACTERISTICS OF DIFFERENT TYPES OF IMC REACTORS

	Packed bed reactors	CSTR	Fluidized bed reactors	Hollow fiber reactors
Ease of operation	Easy	Easy	Difficult	Easy
Ease of catalyst replacement	Difficult	Easy	Easy	Easy
Pressure drop	High	Low	Low	Low
Ease of cell removal	Difficult	Easy	Easy	Easy
Mixing characteristics	Poor	Good	Good	Fair
Oxygen transfer	Poor	Good	Good	Fair
Distribution of viable cells	Uneven	Uniform	Uniform	Uneven
Catalyst attrition	Low	High	Low	Low
Ease of scale-up	Easy	Easy	Difficult	Difficult
Cost	Low	Low	High	Low
Suitability for:				
1. Product inhibited kinetics	Good	Poor	Fair	Good
2. Substrate inhibited kinetics	Poor	Good	Fair	Poor
Ease of reactor control	Difficult	Easy	Easy	Difficult

A. Control of Immobilized Cell Metabolism

As pointed out earlier, it is possible to maintain a high concentration of living cells in the reactor at any dilution rate. This opens up a host of nutrient feed options to impose more sophisticated controls on the immobilized cell metabolism. But the effect of immobilization on the metabolism of microorganisms has not yet been studied in detail. Some evidence of changes in physiology and metabolism has been presented by Navarro and Durand,[22] who observed an increase in respiration rate subsequent to immobilization. However, it is not clear if this increase is due to a shift in metabolism or due to growth (i.e., production of additional biomass) of the cells on the surface.

The physiological state of the immobilized organism would definitely play a role in the rate and amount of metabolite production. For example, in the case of an antibiotic synthesis, the cells may have maturity requirements[23] that would necessitate a carefully controlled growth of the organism. In other words, production of different metabolites would require different strategies for controlling the metabolism of the microbes.

One ready tool available for this purpose is the nutrient composition in the feed. Careful control of the growth factors and inducers in the feed could lead to the maintenance of an immobilized culture in appropriate cell concentration and physiology in order to increase the productivity of the desired metabolite. The control strategy would therefore be to shift the cell metabolism in such a way as to channel more substrate into the desired metabolite. As indicated earlier, it may also be possible to uncouple the growth and product formation phases, i.e., phosphate control of candicidin synthesis. Different options may be available for the practical implementation of the metabolic switch concept; e.g., periodic operation with and without the key nutrient component; continuous addition of the nutrient at a very low concentration. These strategies could be of significant interest in secondary metabolite production.

The other variable at our disposal for controlling cell metabolism is the dilution rate (of course, within the limits imposed by other design considerations). The dilution rate has an impact on the average cell age in the reactor. This has been shown to have an appreciable effect on immobilized cell activity. This approach would be particularly suitable to control the production of growth associated metabolites. Thus, both the nutrient concentration and the feed rate to the reactor would influence the substrate conversion efficiency in the reactor.

B. Cell Growth and Removal Considerations

Under certain conditions there is considerable growth of the cells in both the attached and the bulk phases. This appears to be particularly true for the case of growth-associated product formation. The extent of total growth would be dependent upon the initial cell loading on the matrix, since this effectively acts as the inoculum. In some cases, periodic fluctuations in the product and biomass output have been observed.[15,23] This may result from the changes in the physiology of the organisms. The amplitude of these fluctuations appear to be an inverse function of the dilution rate. The frequency of the fluctuation itself is likely to be a function of the generation time of the organism which, in turn, is affected by the physiological state. These are obviously very interesting observations which need further study and delineation. In passing, it must be mentioned that the dynamics and control of immobilized cell reactions are far simpler in principle than the corresponding chemostat because of the former's inherent stability.[19]

There is also some rudimentary evidence which indicates that the cells preferentially desorb from the matrix upon their death. If this is so, then it could make room for additional replication of the cells within the matrix. The limiting factor here might be the expulsion of the dead cells which is likely to be influenced by the nature of the immobilization matrix and method.

Fungal and mycelial organisms growing in the bulk phase could increase the viscosity of the fluid as well as lead to plugging problems, especially in a packed bed reactor. This may again limit the choice of available reactor types.

C. Substrate Purity

For a single enzyme system, the substrate is usually a clean liquid containing a single component which is easy to handle. For a living cell system, the substrate is likely to be more complex in its composition; it is also likely to contain particulate matter. Some of the problems associated with such complex substrates are foaming, sterility maintenance, and pH control. The last two requirements arise directly due to the use of live cells in the reactor. Therefore, a practical design of the reactor should provide adequate control for these problems. These considerations favor a series of small reactors as they offer better control.

D. Hydraulic Considerations

Particle size and compressibility of the particles is also an important consideration in live cell systems, as they affect the viability of the immobilized cells (since a compressed bed would lead to lesser oxygen availability). Of particular concern is irreversible compaction of the beds. If a fixed bed system is used, we can also envisage plugging problems — especially in the case of growth-associated products. Mycelial fermentation processes are likely to present problems of non-Newtonian rheology and consequent problems of inadequate mixing. The above considerations may impose a maximum allowable cell density in the reactor. It should also be noted that since continuous cell growth assures a theoretically infinite catalyst life, the real life of the reactor would depend on the stability of the carrier matrix.

E. Other Factors

One of the problems often mentioned in continuous culture operation is the possibility of microbial contamination. This problem is reduced to a slight extent in the case of IMC since the cell concentration is usually much higher, but the problem is still a serious one. Simple methods could be used in the laboratory to control this; e.g., sterilization of reactor hardware and feedstock. On a larger scale, it may not be so simple, and no data are available as yet. If sterilization of the nutrient medium could be eliminated in large-scale operation, considerable energy savings could be accrued.

Another problem often encountered in continuous culture is that of culture degeneration due to mutations. Again, this problem may be reduced in IMC systems, but data on this matter are still scanty. Finally, we note that hardly any attempt has been made to scale up the IMC processes for living cell systems. Problems such as oxygen transfer and mixing are likely to affect the scale-up procedures and may dictate changes in reactor configurations.

X. PROCESS ECONOMIC CONSIDERATIONS

Compared to conventional fermentation processes, immobilized living cell systems have the potential to reduce both capital and operating costs. The capital cost reduction stems from the significant increases in volumetric production rate, which lead to dramatically reduced reactor sizes. However, a higher level of sophistication in instrumentation and control would probably be necessary for the efficient operation and control of the reactor systems. This, in turn, would increase the initial capital investment.

Substantial reductions in operating cost can also be expected since continuous operation would result in labor savings. Better process control would result in a more consistent operation and hence some potential savings in material cost. With a reduced initial capital outlay, interest costs and working capital requirements would also be lowered.

Complex bioconversion processes which could be carried out by immobilized living cells fall into two main categories: large volume, relatively low unit value products (e.g., ethanol) and low volume, high unit value products (pharmaceuticals). The impact of IMC technology is likely to be far greater in the former case. Unfortunately, such fermentation processes are often substrate cost intensive. Therefore, the points made earlier regarding substrate conversion efficiency and product concentration in the reactor effluent become even more crucial. Many of the reports in the literature place much emphasis on the reaction part of the overall process. It is propitious to look at the total production and isolation/purification matrix in an integrated fashion. This would point out the economic driving force in the overall process scheme. This problem is addressed in some detail elsewhere.[25]

The greatest potential benefit of immobilization technology for the production of pharmaceutical and other high value products is in the efficient and consistent performance of the conversion process. The use of genetically engineered organisms warrants better containment and better control of the biological processes which may not be achievable in conventional fermentation. Herein lies the greatest opportunity for immobilized cell technology for such products.

The production of several fine chemicals falls somewhere between the two categories mentioned above (e.g., certain amino acids, vitamins). Our own economic analyses indicate that it is often much easier to justify the IMC approach compared with conventional fermentation of chemical synthesis when new production facilities are contemplated.

The economic proposition could be far less attractive in the case of the immobilized system replacing an existing conventional plant. Therefore, the commercial viability of the immobilized cell systems must be evaluated on a case-by-case basis.

XI. EPILOGUE

Immobilized living cell systems offer a great deal of promise in mediating many reaction schemes to produce commercially important products. A few IMC processes have already found their way into industrial practice, albeit the fact that they all involve single enzyme reactions. In this brief analysis we have attempted to provide a broad overview of the process engineering aspects of immobilized living cell systems and to critically appraise the state-of-the-art.

While the technical feasibility of employing the immobilized living cell concept has been demonstrated for a number of cases, fundamental analysis of the IMC catalysis is lacking. In particular, much needs to be learned about the physiology and energetics of the microorganisms in the immobilized state. Many engineering problems also remain to be solved.

Further developments in the field should steer toward this direction so that a sound basis can be formulated to assess the realistic potential of immobilized living cell systems.

ACKNOWLEDGMENTS

The authors wish to thank Mrs. Terri Kumpa for typing and proofreading the manuscript.

Parts of this work were supported by the National Science Foundation (Grant CBE-80-10865) and the H. J. Heinz Company, for which the authors are grateful.

REFERENCES

1. **Abbot, B. J.**, Immobilized cells, in *Annual Reports on Fermentation Processes*, Vol. 1, Perlman, D., Ed., Academic Press, New York, 1977, 205.
2. **Abbot, B. J.**, Immobilized cells, in *Annual Reports on Fermentation Processes*, Vol. 2, Perlman, D., Ed., Academic Press, New York, 1978, 91.
3. **Jack, T. R. and Zajic, J. E.**, Immobilization of whole cells, in *Advances in Biochemical Engineering*, Vol. 5, Ghose, T. K., Fiechter, A., and Blakebrough, N., Eds., Springer Verlag, Berlin, 1977, 125.
4. **Chibata, I.**, Enzyme engineering in food processing, in *Food Process Engineering*, Vol. 2, Linko, P. and Larinkari, J., Eds., App. Sci. Pub., London, 1980, 1.
5. **Venkatasubramanian, K., Ed.**, *Immobilized Microbial Cells*, ACS Symp. Ser. 106, American Chemical Society, Washington, D.C., 1979.
6. **Venkatasubramanian, K. and Vieth, W. R.**, Immobilized microbial cells, *Progr. Ind. Microb.*, 15, 61, 1979.
7. **Constantinides, A., Bhatia, D., and Vieth, W. R.**, Immobilization of *Brevibacterium flavum* cells on collagen for the production of glutamic acid in a recycle reactor, *Biotechnol. Bioeng.*, 23, 899, 1981.
8. **Larreta Garde, V., Thomasset, B., and Barbotin, J. N.**, Electron microscopic evidence of an immobilized living cell system, *Enzyme Microb. Technol.*, 3, 216, 1981.
9. **Hattori, R., Hattori, T., and Furusaka, C.**, Growth of bacteria on the surface of anion exchange resin, *J. Gen. App. Microbiol.*, 18, 271, 1972.
10. **Messing, R. A., Opperman, R. A., and Kolot, F. B.**, Pore dimensions for accumulating biomass, in *Immobilized Microbial Cells*, ACS Symp. Ser. 106, Venkatasubramanian, K., Ed., American Chemical Society, Washington, D.C., 1979, 12.
11. **Messing, R. A.**, Support-bound microbial cells, *Appl. Biochem. Biotechnol.*, 6, 167, 1981.
12. **Chibata, I. and Tosa, T.**, Industrial applications of immobilized enzymes and immobilized cells, in *Applied Biochemistry and Bioengineering*, Vol. 1, Wingard, L. B., Jr., Goldstein, L., and Katchalski, I., Eds., Academic Press, New York, 1976, 118.
13. **Mattiasson, B., Anderson, E., Albertson, P. A., and Hahn-Hägerdal, B.**, Bioconversion in aqueous two-phase systems, in Proc. 6th Enzyme Engineering Conference, September 20-25, 1981, Kashikojima, Japan.
14. **Karkare, S. B. and Venkatasubramanian, K.**, Unpublished results, 1981, Dept. of Chem. and Biochem. Eng., Rutgers Univ., New Brunswick, N.J.
15. **Bandyopadhyay, K. K. and Ghose, T. K.**, Studies on immobilized *Saccharomyces cerevisiae*. III. Physiology of growth and metabolism on various supports, *Biotechnol. Bioeng.*, 24, 805, 1982.

16. **Vieth, W. R. and Venkatasubramanian, K.**, Immobilized cell systems, in *Enzyme Engineering*, Vol. 4, Broun, G., Manecke, G., and Wingard, L. B., Jr., Eds., Plenum Press, New York, 1978, 307.
17. **Tyagi, R. D. and Ghose, T. K.**, Studies on immobilized *Saccharomyces cerevisiae*. I. Analysis of continuous rapid ethanol fermentation in immobilized cell reactor, *Biotechnol. Bioeng.*, 24, 781, 1982.
18. **Liu, C. M., McDaniel, L. E., and Schaffner, C. P.**, Factors affecting the production of candicidin, *Antimicrob. Agents and Chemother.*, 7, 196, 1975.
19. **Venkatasubramanian, K., Karkare, S. B., and Vieth, W. R.**, Chemical engineering analysis of immobilized cell systems, in *Applied Biochemistry and Bioengineering*, Wingard, L. B., Jr. and Chibata, I., Eds., in press.
20. **Venkatasubramanian, K. and Harrow, L. S.**, Design and operation of a commercial immobilized enzyme reactor system, *Ann. N.Y. Acad. Sci.*, 326, 141, 1979.
21. **Venkatasubramanian, K.**, High fructose corn syrup: plant design and process economics, in *Enzymes: The Interface Between Technology and Economics*, Danehy, J. and Wolnak, B., Eds., Plenum Press, New York, 1980, 34.
22. **Navarro, J. M. and Durand, G.**, Modification of yeast metabolism by immobilization onto porous glass, *Eur. J. Appl. Microbiol.*, 4, 243, 1977.
23. **Martin, J. F. and McDaniel, L. E.**, Kinetics of biosynthesis of polyene macrolide antibiotics in batch cultures: cell maturation time, *Biotechnol. Bioeng.*, 17, 925, 1975.
24. **Hattori, R.**, Growth of *Escherichia coli* on the surface of an anionexchange resin in continuous flow system, *J. Gen. Appl. Microbiol.*, 18, 319, 1972.
25. **Venkatasubramanian, K.**, Integration of large scale production and purification of biomolecules, in *Enzyme Engineering*, Vol. 6, in press.

INDEX

A

B

C

D

E

F

G

H

I

N

O

P

R

S

T

For Product Safety Concerns and Information please contact our EU
representative GPSR@taylorandfrancis.com
Taylor & Francis Verlag GmbH, Kaufingerstraße 24, 80331 München, Germany

www.ingramcontent.com/pod-product-compliance
Ingram Content Group UK Ltd.
Pitfield, Milton Keynes, MK11 3LW, UK
UKHW052221190926
475702UK00006B/1148
* 9 7 8 1 3 1 5 8 9 4 2 8 7 *